On Dragons' Wings

I0895969

In loving memory of my mother-in-law,
Brenda West, who passed away in the
final stages of this book's production.

On Dragons' Wings

A History of No. 614 (County of Glamorgan) Squadron, Royal Auxiliary Air Force

MARTIN WADE

y lolfa

All proceeds from the sale of this book will be donated to the Royal Air Force Benevolent Fund and the South Wales Aviation Museum

Cover photo: Pilots return to RAF Llandow, *c.*1948, with Spitfire MkXVI behind.
Peter Irving-Lewis

Back cover photo: The detachment from 614 Sqn deployed to run a mobile testing unit during the pandemic in 2020. Back, from left: SAC Jon Bridge, SAC Hugh Glyn-Jones, LAC Jack Pritchard, SAC James Rowlands-Jones, SAC Richard Steer; Middle row, from left: SAC Bradley Fletcher, SAC Hannah Young, Cpl Kate Tatchell, LAC Michelle Lewis; Front row, from left: Sgt Cathy Sharples, Cpl Gareth Jones.
Sgt Cathy Sharples

First impression: 2024

Cover design: Y Lolfa

ISBN: 978-1-80099-599-4

Published and printed in Wales
on paper from well-maintained forests by
Y Lolfa Cyf., Talybont, Ceredigion SY24 5HE
website www.ylolfa.com
e-mail ylolfa@ylolfa.com
tel 01970 832 304

Foreword

In this centenary year of the Royal Auxiliary Air Force, it is wonderful to welcome this history of No. 614 (County of Glamorgan) Squadron, the first dedicated history of Wales' only RAuxAF Squadron.

Over many years, the RAuxAF have served our country with loyalty and bravery, making an important contribution to our nation's security and defence, particularly during the many periods of war and crisis of both the last century and the current one. Today, the RAuxAF very much continues to be an important part of the RAF's order of battle, staffed by skilled and enthusiastic people, who lead very busy lives, holding down a civilian job on top of their RAuxAF service. They are truly our modern-day citizen Air Force.

In my recent RAF role as Air Officer Wales, which is the senior RAF position in Wales, I often used to highlight to the Welsh public the incredible history of service that countless Welsh men and women have given to the RAF over many years, together with the support Welsh communities have given to RAF stations located here in Wales. In doing that, I always viewed what the RAF achieved, and continues to achieve today, as a great team effort and it was always wonderful to witness both the interest and the pride of Welsh people in the RAF and the contribution Welsh men and women have made, and still do today.

On Dragons' Wings is a wonderful addition to that rich history of the RAF and Wales, telling the story of No. 614 Squadron from its formation in the 1930s, flying biplanes, to its extensive service throughout the Second World War, to flying Vampire jets out of RAF Llandow in the Vale of Glamorgan. The story takes us to all corners of the globe, but it is very fitting that it finishes with the 614 Squadron of today, operating out of the Welsh capital and, whilst no longer flying aircraft, still carrying out a range of important tasks for the defence of the United Kingdom. Importantly, *On Dragons' Wings* also tells the stories of the many people that served in the squadron, the service they gave, the many that lost their lives, together with a healthy smattering of the many more light-hearted moments and escapades that also form part of life in the military.

I hope you enjoy this fascinating book. It has much to commend it, including 614 Squadron's wonderful Welsh motto, *Codaf i Geisio* (I rise to Search/Seek).

Air Commodore Adrian Williams OBE RAF (Rt'd)

Preface

It is December 1942. In the clear skies over Tunisia, nine Bristol Bisley bombers, flying to attack an enemy airfield, turn and descend to their target. Suddenly there are shouts over the radio. Fighters are spotted. Quickly the sky is black with Me109s. Like a swarm, they come from all directions attacking them. It is the beginning of a massacre.

One of the bombers under attack is flown by Bill Irving of No. 614 Squadron. The pilot from Cwmbran is flying number three in the leader's formation. He breaks away, trying to evade three Me109s. Flying at almost ground level down along a valley floor, he desperately jinks his lumbering Bisley trying to avoid the fire pouring down from the German fighters. His rear gunner hits one of them which rolls on its back and brakes off the chase, smoke pouring from its belly before his guns jam. Irving's port engine and starboard aileron are hit, but he continues to nurse his one-engined bomber for another 15 minutes before belly-landing in the desert. He and his crew escape unharmed and walk to Allied lines. All nine of the attackers are shot down with few survivors. After the war Irving would return to Wales to become the squadron's Commanding Officer in 1946.

His bomber had been armed that day by groundcrew at Canrobert further west in Algeria. Among them was Sgt Bill Hooper, from Canton in Cardiff. He joined 614 Squadron as a reservist in 1937 when the squadron formed at Pengam Moors on the outskirts of the city. He'd continued to work as a mechanic in civilian life but served as an Aircraft Fitter in his spare time. Alongside 160 others, he had been mobilised at the outbreak of war and together they marched out of Pengam Moors to war in 1939. Their numbers had been swelled by others recruited from across the UK and the Commonwealth and, though the squadron would change theatres and roles – guarding against German invasion in the UK; helping defeat Rommel's forces in North Africa, then to Italy and the skies over the Balkans and Eastern Europe as an elite Pathfinder unit – its identity would remain as Glamorgan's squadron. Its Welsh-language motto 'Codaf i Geisio' (I rise to search or seek) offered a hint of its many and varied roles. When the war ended the squadron returned to Wales and flew Spitfires and then Vampire jets from Llandow as a Cold War fighter squadron before it disbanded, along with much of the Royal Auxiliary Air Force, in 1957. It reformed again in 2014 and its members now serve the RAF across the world.

This is the story of those reservists from Wales, but also those who came from Scotland, England, Ireland, Guernsey, Canada, New Zealand, Australia and South Africa to serve in 614 – and the reservists of today who serve in what is still the only RAF Reserve Squadron in Wales.

It is fitting that this book is published on the 100th anniversary of the Royal Auxiliary Air Force and is about its Welsh unit. As such, its story is unique and it is a story that has never been fully told.

It is a story of machines, training, equipment, risk, danger and sacrifice – sometimes the ultimate sacrifice – but always, then and now, it is a story of people. What brings them together and makes them able to work hard, become ready, confront danger and rely on each other.

My main job for the RAF is as a media operations officer, and I've been fortunate to have covered many deployments of the RAF around the world. It's been my privilege to report on the people who serve in the RAF today – from pilots to chefs – and to ensure that their voices are heard. Similarly, working on my corresponding role as a journalist in civilian life, a great pleasure and honour was writing historical features for my daily paper, the *South Wales Argus*, often on military themes. It was a privilege too in connecting with the past, to allow unheard voices to be heard and give them their say. Even if often from beyond the grave, I felt it was important that I did them justice and that I honoured their memory.

It is my earnest hope that this book – while longer than those features – does the same thing and tells their stories and honours their memories.

Acknowledgements

TO THE RECENTLY retired Air Officer Wales (AOW), Air Cdre Adrian Williams, I thank you for writing the Foreword to this book. It was a pleasure to work with you in helping to tell the story of the many links between Wales and the RAF, and I believe those links are stronger as a result of your hard work and dedication.

To those of you it's been my privilege to serve with, I give you my sincere thanks. To our former CO, Wg Cdr Phil Sagar, who led the reformation of 614 and opened the door for me. To my media boss, Wg Cdr Andy Pawsey, for being so generous with encouragement and trust. The ever-capable Media Flight, who it's my privilege to lead, and to the fantastic men and women of 614 Squadron, from intelligence analysts to MT drivers, who it's my honour to serve alongside. To our current CO, Wg Cdr Olly Walker, I thank you for your leadership as we move into an exciting future and support, not least of my ideas to ensure that 614 Squadron's story is heard.

It's for us, as squadron members, to walk in the footsteps of those who once served in 614 and whose stories make up this book. Most of them are no longer with us and some lie in faraway fields. I salute them all and the sacrifices they made. It is to their memories and to those who serve in 614 in modern times that this book is dedicated.

I thank the relatives of our veterans who have given so generously of their stories, pictures, memories and advice, in particular Wendy Barnes, John Bonner, Nick Budd, Diane Crawley, Judy Cleasby, Mike Fleckney, Ian Hooper, Peter

Le Cheminant, Robin Irving-Lewis, Jo Maddaford, Hamish MacLaren, Michelle McWilliams, Billie Moore, Helen Mussell, Peter Scotland, Craig Shilliday, Dermot Stapleton and Alison Patriarca. Also, Caroline Whitfield for keeping the flame of remembrance alive through the Squadron Association. This book could not have been written without you.

Lastly, I thank my wife Su and daughter Carys. Countless are the nights when you tolerated me tapping away on my laptop, distant and oblivious to your cares and questions. You have put up with and supported this labour of love and your support is repayed with love from me.

Diolch o galon i chi gyd.

Martin Wade
May 2024

Contents

List of Abbreviations 15

Chapter 1: Formation (1937–1940) 17

Creation at the cradle of aviation in Wales; the squadron is born; learning to fly; the coming of war; 'Phoney War'

Chapter 2: Scotland (1940–1942) 46

Desperate preparations against invasion; chemical weapons and pick-axe handles to fight Nazi invaders

Chapter 3: Raids (1942) 74

Thousand-Bomber Raids; Dieppe Raids

Chapter 4: North Africa (1942–1943) 102

Journey south; 'Operation Torch' and arrival in Algeria; first raids; a posthumous VC is awarded amid heavy losses; disaster averted at the Kasserine Pass; the final push and defeat of the Axis in North Africa

Chapter 5: Coastal Patrol (1943–1944) 137

Farewell to Africa; lonely Mediterranean search and rescue missions

Chapter 6: Italy, early days (1944) 158

Learning the Pathfinder art; early missions and disastrous raid on Bulgaria

Chapter 7: Italy, oil (1944) 188

Strikes on vital heavily-defended oil refineries – many are lost

Chapter 8: Italy, POWs (1944–1945) 205

Shot down and captured; prisoner-of-war life in Romania, Hungary and Nazi Germany; surviving the death march across a frozen Reich

Chapter 9: Italy, the road to victory (1944–1945) 233

Targeting transportation; strikes on bridges, railways and harbours; parachuting supplies for Yugoslav partisans in the Balkan fight; precision air strikes for the Army; VE Day, taking POWs home and disbandment

Chapter 10: Llandow (1946–1957) 267

Reformation and return to Wales; Llandow air disaster and the death of the CO; Cold War mission to stop Soviet bombers; jet age; summer camps across Europe; disbandment

Chapter 11: Interregnum (1957–2014) 287

The founding of the 614 Association; the gathering of flags, memories and memorials

Chapter 12: 614 Reformation (2014 to the present) 292

Rebirth and new role; supporting RAF operations in Wales, the UK and across the world, including the pandemic

Endnotes 311

Selected Bibliography 330

Index 334

List of Abbreviations

AAF: Auxiliary Air Force
AC, AC1, AC2: Aircraftman/woman, 1st Class, 2nd Class
Air Cdre: Air Commodore
AM: Air Marshal
AOC: Air Officer Commanding
AS1, AS2: Air Specialist Class 1, Class 2
ASR: Air Sea Rescue
AVM: Air Vice Marshal
Bdr: Bombardier (Bomb-Aimer)
Brig: Brigadier
Btn: Battalion
CO: Commanding Officer/OC: Officer Commanding – both mean the same thing, i.e. the commander of the squadron, but both are used
Cpl: Corporal
F/Eng: Flight Engineer
F/O: Flying Officer
F/S: Flight Sergeant
Flt Cdr: Flight Commander
Flt Lt: Flight Lieutenant
Flt: Flight
GEE: Radio-navigation system used by the RAF in the Second World War
Gen: General
Grp Capt: Group Captain
Grp: Group
H2S: Ground-mapping radar
HCU: Heavy Conversion Unit
JNCO: Junior Non-Commissioned Officer (Corporal)

LAC: Leading Aircraftman/woman
Lt Gen: Lieutenant General
Lt: Lieutenant
MAAF: Mediterranean Allied Air Force
MACAF: Mediterranean Coastal Air Force
MAF: Mediterranean Air Forces
MASAF: Mediterranean Allied Strategic Air Force
Obs: Observer
OTU: Operational Training Unit
P/O: Pilot Officer
PFF: Pathfinder Force
RAAF: Royal Australian Air Force
RAF: Royal Air Force
RAuxAF: Royal Auxiliary Air Force
RCAF: Royal Canadian Air Force
RNZAF: Royal New Zealand Air Force
SAAF: South African Air Force
SAC: Senior Aircraftman/woman
Sgt: Sergeant
SNCO: Senior Non-Commissioned Officer (Sergeant, Flight Sergeant, Warrant Officer)
Sqn Ldr: Squadron Leader
Sqn: Squadron
T/Gnr: Tail Gunner
TBF: Tactical Bomber Force (North Africa)
TI: Target Indicator – flares dropped to mark targets
TM: Target-Marking
TMB: Target-Marking Bomb
U/T: Under Training
USAAF: United States Army Air Force
W/O, W/O 1: Warrant Officer, Warrant Officer, 1st Class
Wg Cdr: Wing Commander
Wg: Wing
WOP/AG: Wireless Operator/Air Gunner
WOP: Wireless Operator

CHAPTER 1

Formation (1937–1940)

PENGAM DOESN'T LOOK like promising ground for flight. Edged by the mud of the River Severn and the River Rhymney, it sits on marshy land on the old eastern border of Glamorganshire. Its name comes from the Welsh, 'Pen' (head or end) and 'Cam' (crooked or bent). This bend is the Rhymney's last meander before it reaches the sea.[1]

Number 614 (County of Glamorgan) Squadron, Auxiliary Air Force, was created at the fringe of what was then the county's only city and largest settlement – Cardiff. The airfield from where it first took flight was known by many names and the site of its first airfield lies in a place with many labels. There is Pengam, there is Tremorfa, and the place sits within a larger district rejoicing in the name of Splott. Its airfield was first known as Splott Aerodrome, then Cardiff Municipal Airport, then RAF Pengam Moors, then RAF Cardiff. This windswept place, by whatever name it is called, was where it all began on 1 June 1937. But other things had taken off from here long before then.

Far-seeing pioneers brought flight here early. One who saw its potential was Ernest Willows. He used this unpromising looking marsh land at the edge of fast-growing Cardiff as the launch pad for his early forays into the air, flying airships from here in the early 20th century. Born on nearby Newport

Road in Cardiff, he dazzled the Edwardian city with his exploits, flying in his airship over the city in June 1910 to land outside Cardiff's new City Hall, watched by a huge crowd.

His name is remembered by the school Willows High School, which was built on the site from where his early launches took place and from where 614 Squadron would be created eleven years after his death.[2]

Keen for the Welsh capital to connect with the growth of air travel, in 1931 Cardiff City Council embarked on a scheme to build an aerodrome. As Willows had found, despite first impressions, Pengam was actually well-suited for flying, with approaches over the sea avoiding hills and housing. The area, however, was vulnerable to flooding and £18,000 was allocated to build a sea wall, drain and level the site.

Despite continuous rain waterlogging the airfield that winter, in April 1932, Splott Aerodrome received its Air Ministry licence. Scheduled air services began flying to Birmingham, Plymouth or Weston-super-Mare in biplane airliners.

In the few years leading up to the squadron's formation in 1937, there was increasing interest from the Air Ministry in south Wales. With the Nazis seizing power in Germany in 1933, planning for possible war gathered pace.

Between 1934 and 1939 there were numerous expansion plans for the RAF which called for an enormous increase in strength and re-equipment with a new generation of aircraft like the Hurricane and Spitfire.

These plans also saw the establishment of Bomber, Fighter, Coastal and Training Commands in 1936, followed by Maintenance Command in 1938 and Reserve Command in 1939. Bomber and Fighter Commands would be deployed in the east of the country to face any potential aggressor. Training and maintenance units would be placed in the west where they would be safe from German air attack. This made Wales an attractive location for them.

Before 614 Squadron was formed, an older reserve unit already existed across the Bristol Channel and they became regular visitors to Splott Aerodrome. Filton-based No. 501 (County of Gloucester) Squadron often used the new airfield, and in July 1934 Cardiff Aeroplane Club and the Deputy Lord Mayor hosted the crews of twelve of their Westland Wallace biplanes.[3]

The squadron was formed in 1929 and was an early sign of the growth in the RAF Reserves. Plans for this part-time air force had been laid when the RAF was created in 1918 and the first Reserve Squadrons were formed in 1925.

The concept of reserves, or a part-time military force, had long existed in Britain, but in 1907 local militias and volunteer units were replaced with a system of territorial regiments raised by county associations. This was the system that squadrons like 614 would later be based on.[4]

The Auxiliary Air Force and Air Force Reserve Bill was laid before the House of Commons in March 1924. It extended the provisions of the Territorial and Reserve Forces Act of 1907 to what was to be called the Auxiliary Air Force (AAF). This formed the foundations of squadrons like 614.

The first squadron to be formed was No. 502 (Ulster) Squadron on 15 May 1925. Nos 602 (City of Glasgow), 600 (City of London), 601 (County of London) and 603 (City of Edinburgh) followed later that year. More squadrons were formed in the 1920s and early 1930s. Nos 503 (City of Lincoln) and 605 (County of Warwick) in 1926, 504 (County of Nottingham) in 1928, 501 (County of Gloucester) in 1929. Nos 604 (County of Middlesex), 607 (County of Durham) and 608 (North Riding) in 1930. No. 500 (County of Kent) followed in 1931. By 1933 the AAF had 1,335 personnel.[5]

With the worsening political situation in Europe in the mid-1930s, expansion of the RAF accelerated, with a further eight AAF squadrons created.

Of the new AAF squadrons, three formed in 1936. Nos

609 (West Riding), 610 (County of Chester) and 611 (West Lancashire).

Following speculation in the autumn of 1936, the Air Ministry began to reveal its plans for south Wales. Within the space of a few months it was announced that one of four new Auxiliary Air Force squadrons was to be based at a considerably enlarged Pengam.[6]

While all previous Auxiliary and Reserve units were light bomber or fighter units, three more formed in 1937 were given the job of Army Co-operation. They were Nos 612 (County of Aberdeen), 615 (County of Surrey) and 614 (County of Glamorgan) at Pengam. No. 503 (City of Lincoln) Squadron was disbanded, to reform in 1938 as 616 (South Yorkshire) Squadron and No. 613 (City of Manchester) was the last to form in 1939.

On 1 June 1937, No. 614 Squadron's founding made the news. Under the headline 'Cardiff Air Auxiliary Squadron', the *Western Mail* reported that, 'To-day is the official day for the establishment of the first Auxiliary Air squadron in Wales.' Readers were told that, 'Already there have been more than 30 applications for commissions. The squadron is being formed in Cardiff, and will be known as the No. 614 County of Glamorgan Army Co-operation Squadron, Auxiliary Air Force.'

A commanding officer had not yet been appointed, but the names of 'certain men' have been forwarded to the Air Ministry, and the successful applicant would be known within the next few days.[7]

Some, though, had already joined. An advance party of regular personnel arrived some months before the recruitment of part-time reservists. Its first adjutant, Flt Lt P Gilbert Morgan-Smith, opened an office at what was by now called Cardiff Municipal Airport at Pengam. The squadron also had an office at Dumfries Place in the centre of Cardiff, which it shared with the Territorial Army. Joining this party was, in fact, a reservist, and so was the very first part-timer recruited to 614. Dr Ieuan Rhys-Jones was the squadron's first Medical Officer and was transferred from the Territorial Army to care for those RAF regulars.

On 15 June the first aeroplanes, three Avro Tutor trainers, arrived and were housed in the Cardiff Aeroplane Club hangar at Pengam. Flying training for the first recruits started at the beginning of August.[8]

One of these regulars was Cpl Edgar Salmoni, an aircraft mechanic from Newport, who would stay with 614 until 1942. Alongside other airmen were two instructor pilots, F/Lt Morgan-Smith, the adjutant, and F/Lt Thomas-Ferrand.

Edgar remembered when they received their first part-time recruits. 'Seven or eight civilians came into our half of the hangar. One was Charles Cadman, the son of Richard Cadman, who would become squadron CO. Another was Bill Drummond and a Mr Llewellyn. These early entrants were christened the "818s" – so-called because their service numbers all began with the number "818".'[9]

Bill Drummond was recruited by curious means. His friend and fellow engineer, John Rollinson, who worked on Penarth Road in Cardiff, had flown from Pengam as a passenger on a Cambrian Airways flight. There they'd met the adjutant, Flt Lt Morgan-Smith.

'A few weeks later John and I were in the barbers at the Angel Hotel [in Cardiff], when in walked Morgan-Smith. He told John that lectures on navigation etc., would soon be started by the squadron at the airport, and hoped he would come along. He turned to me and said he hoped I would come also.

'In the event, I was commissioned on 2 May 1938, and John, I think, on the following day. I had gone solo on the 10th of April 1938 and John had beaten me by a few days.'[10]

Another of the first recruits, or '818s', was recruited by more conventional means. In 1938, Hector MacDonald attended a recruitment open day for the territorial forces at the drill hall on Dumfries Place. One of the units taking part, he recalled, was 614 Squadron. 'I put my name down and, early in 1939, I found myself one of the second entry, 818088, AC2 (Aircraftman 2nd Class) U/T (under training) Wireless Operator.'[11]

There were many motivations for joining the new squadron. Another '818', Cliff Haines, was attracted by the prospect of training and work in a time of high unemployment. 'I was a young man who left school at the age of 14 with a very limited education. Britain was just coming out of that long depression of the 1930s, leaving you to take whatever was on offer. I was always interested in engineering, so when I heard that an Auxiliary Air Force unit was to be formed at Pengam Moor and that they were taking on recruits, I applied, was accepted and given a number – 818206.'[12]

The money on offer to new recruits was a real attraction. For some, like Cliff Haines, it represented their sole income, for others like Tom Coles it was a welcome supplement to a poorly-paid day job. He remembered 'those heady days of 1938 and 1939', training on Monday and Thursday nights and all day on Saturday and Sunday, paid at 1*s.*3*d.* per hour. 'It certainly supplemented my weekly wage of 17*s.*6*d.* from the *Western Mail & Echo*,' he recalled.[13]

A later feature in the paper that Cliff worked for, the *South Wales Echo*, told how the pilots were paid an extra shilling for every flying hour, so at weekends many of them stayed up from dawn to dusk – with a break for lunch.

Dai Park was inspired to join the squadron by a kindly RAF Sergeant who captured the Cardiff teen's imagination at a Pengam airshow before 614 formed. There, he explained the mysteries of a Handley Page Heyford bomber to the youngster. 'I remember climbing through the aircraft and being fascinated by it and that was the spark that made me determined to become an aircraft fitter and eventually to be a member of 614 Squadron.'[14]

Some recruits were already experienced in their trades and Bill Hooper, a mechanic from the Canton area of Cardiff, found out the hard way that unsought advice to his regular colleagues was not always appreciated. He and fellow recruit AC2 FA Baker were exploring the Pengam camp on their first Sunday in uniform.

'We came to the east hangar and looked inside.' They saw a sleek silver Hawker Hind with its tail on a trestle. Some regulars, under a corporal, were heaving and struggling to remove its Kestrel engine. It was, he recalled, 'an education in strong language and misapplied energy'. He looked at the Coles crane and noted the position of the jib. 'It was too near the aircraft and the jib head was too far back so that the angle of the wire was working against them.'

The experienced technician thought he could help. 'Excuse me, corporal,' he said. 'If you move the crane forward a few feet, the weight of the engine will work for you instead of against you.' The corporal exploded. '"Bloody civilians!" he shouted, "they haven't got the creases out of their uniforms and already they're trying to tell us our job. Bugger off with you!"

'We beat a hasty retreat,' he recalled, 'but I looked back and they were moving the crane!' It was, he remembered 'a bad start', but added, 'many of us were time-served mechanics already'.[15]

There were still bumps to be ironed out before those early recruits were fully accepted. Edgar Salmoni remembered the Saturday morning when the first uniform issue took place for the Auxiliaries. 'Some had trousers too long, others too short, some tunics too small, some too big, some with ties, others without. One had his plus-fours on with his tunic, another no black tie but a bright spotted one – the tailor had a field day making alterations!'

One of the regulars was a F/Sgt Pope. Edgar remembered him complaining, 'I hope we don't have a war. I've heard of Fred Karno's Army, but this lot is the limit.' And yet, Edgar recalled, within a week the fitting was put right and the pride of belonging to the squadron began to build.[16]

That pride was there in recruits regardless of rank and background. Yet there is a sense that 614 Squadron was a little different to other AAF squadrons.

The vision of the RAF's founder, Air Marshal Sir Hugh Trenchard, was that a fighting service must possess a non-regular branch with its roots firmly set in the civilian life of the country. He saw his Auxiliaries as modern successors of the old mounted yeomanry. Instead of riding horses, these men who were now 'attracted to flying machines, who had leisure and a thirst for adventure, were prepared to indulge in uniform.'

The social background of the AAF reflected this and can be seen by the leaders of the squadrons already formed. Known as the 'the Millionaires' Squadron', No. 601 (County of London) had as its first CO Lord Edmond Grosvenor, and its distinguished membership included Sir Philip Sassoon who was later to become Under-Secretary of State for Air. The Marquis of Clydesdale flew with No. 602; Lord Willoughby de Broke with No. 605 and Viscount Runciman with No. 607.[17]

Lieutenant Colonel Dore, CO of 604 (County of Middlesex) Squadron, gave marks for school record in scholarship and athletics. Extra marks were given if the potential officer rode a horse, drove a car or a motorcycle, could sail a boat, ski or play the piano.[18]

The Welsh squadron seemed to be different. Its first CO was Sqn Ldr Richard Cadman. A Royal Flying Corps captain in the First World War, he was a prominent south Wales businessman and director of Cardiff Aeroplane Club. While he might be described as middle-class, he was clearly not one of the 'millionaires' who dominated many of the English squadrons.

One of 614's earliest part-time pilots, Richard Rhys, recalled how Cadman set out his stall. 'There were about twelve or fifteen of us, who had learned to fly in university squadrons or in flying clubs. But Dick Cadman said it wasn't going to be a rich man's squadron. He made it a rule that anyone with more than £50 a year unearned income couldn't join.'[19] Nevertheless, many were from professional backgrounds, albeit in the early stages of their careers. One, Martin Edwards, joined as a pilot in 1937 and worked at his family's law practices in Merthyr Tydfil and Cardiff after graduating from Oxford. Another was chartered accountant Glen Pallot, who worked on Bute Street in Cardiff.[20]

Although 614 differed from their mostly English counterparts, some of their activities chimed with the club-like atmosphere (for the officers at least) in the Territorial tradition – complete with what might be called 'japes'. Rhys recalled, 'We used to do mad things. We'd fly to Paris for a Saturday night out and I remember one party where the officers' bedroom doors were sawn in half so the place looked like a stable. Other times we might pop up to lunch at Hendon with 604 Squadron.'[21]

In those early days, Flt Lt Thomas-Ferrand would fly Edgar Salmoni in one of the Avro Tutors to the neighbouring AAF unit, No. 501 Squadron, in Bristol. Edgar recalled, 'I was partly used as ballast, but he would meet with their officers and I would go to their workshops to make something or other or pick up a few tips on what was needed to stock up the metal-workers' shop.'[22]

As well as the Tutors, the squadron also acquired a single Hawker Hart and two Hawker Hinds for training, with the first delivered in January 1938.[23]

An early recruit and one of the first part-time pilots to earn his wings with 614 was P/O Peter Farr, who had joined in

1937. 'By March 15th 1938, I had already completed my initial training on Avro Tutors with the squadron, mostly instructed by Hugo Thomas-Ferrand. We had obtained an old Hawker Hart two-seater trainer on which I had gone solo that very month.' He was told at the time, 'Don't bust it – it's the only one we've got.'[24]

The squadron's first operational type was the Hawker Hector and the first of them was received on 12 April 1938.[25] For many newly-qualified pilots, the Hector was far more challenging to fly than the Hart or the Hind. It was one of the latest types the RAF had acquired and had only entered service in 1937.

The smooth lines of the Hart's Kestrel engine were replaced by the Napier Dagger, an air-cooled engine with its 24 cylinders arranged in an unusual 'H' cross-section. It gave 805hp – a jump from the 525hp of the Hart, and the swept-back upper wing, so characteristic of the Hart series, gave way to a straight wing.[26]

A maths student at University College, Cardiff, Peter Farr recalled a fine spring day when, on a whim, he decided to skip 'boring maths lectures' and headed for Pengam Moors where he had his first flight in a Hector.

He remembered, 'I turned in through the white barred aerodrome gate, along the cinder path, when suddenly I saw a beautiful silver aircraft standing on the grass and facing away from me. As I stood gazing at this, a heavy arm went around my shoulders and a voice said, "Do you understand the fuel system?" Of course, I didn't have a clue, so naturally I said "yes", and P/O Norman Merrett, Flight Commander "B" Flight said "OK – off you go."

'So, collecting my helmet, goggles, parachute and gloves, I climbed into the cockpit.'

The gleaming Hector was an impressive machine – its14-foot propeller thrust skywards atop its brutish-looking engine. He recalled how it was started by two groundcrew turning starter handles, slowly at first, then with a mounting whine from the flywheel, they went faster and faster. This 'inertia starter' was very effective he remembered. 'Once up to speed, with throttle set and switches on, a toggle under the fuselage was switched to spin the engine while the pilot frantically turned the hand starter magneto in the cockpit.' After this manic activity, he recalled 'the engine almost invariably fired' and the 24-cylinders would roar into life.

'The day was fair with light airs from the south – which was the shortest width of the aerodrome,' Peter recalled. Once vital checks were done, he gave full throttle and pulled the boost lever back. Fuel would course through the line to feed the roaring engine and the silver brute would strain against its brakes.

Taking off over the Severn estuary, Peter soon realised what kind of machine he was flying. 'Once airborne it became obvious that this was a "hot rod" aircraft compared with the Hart trainer.' The Hector's top speed was 187mph, and although the RAF's latest fighter, the Hurricane, entered service that year with a top speed more than double that, the Hector was clearly still a sprightly performer compared to the squadron's existing Tutors and Hart.[27]

Landing could be tricky and he recalled that some pilots nosed over after touching down and many came to grief when they converted to the Hector. On 3 April 1938, a Hector K9784 was damaged beyond repair and its pilot, P/O Richard Rhys, was injured after it hit a bump on landing at Pengam and the machine tipped forward. A similar fate befell K8109 on 4 June after the pilot braked heavily after landing and it overturned.[28] Further from home, K9728 was damaged on 7 September 1938 after force-landing in bad weather on a navigation exercise. The crew, P/Os John Rollinson and Martin Edwards, were flying from Glasgow when they encountered fog six miles from their destination at RAF Sealand. They were lost for two hours before putting down in a field at Oakenshot Farm in Flintshire, but overshot into a hedge.[29]

As with Edgar Salmoni's early jaunt in the Tutor, groundcrew also got to fly. Fitter Bill Hooper was working on an aircraft starter motor when the pilot, Glen Pallot, an early 614 flyer, came out to test-fly it. 'He asked if he wanted a flight and I eagerly accepted.'

The corporal he was working with explained to Bill how to load the parachute. 'He told me to pull it through the side hatch and onto the floor of the rear cockpit. Unfortunately, the release handle of the 'chute caught on something and the cockpit filled with silk!'

'Ashamed and mortified', Bill was told to gather the billowing unfurled cloth, carry it and run 500 yards back to the parachute section and draw another from the store. Winded from his unexpected dash across the airfield, Bill clambered onboard. They flew along the Glamorgan coast, taking in Llantwit Major, around 25 miles west of Pengam. Bill recalled, 'My stomach stood up pretty well and we were coming back to the aerodrome and flying with my back over Roath Park Recreation Ground a few miles north of Pengam.' By this time Bill had begun to feel quite ill and this discovery that he was prone to motion sickness meant he tried to avoid flying in future.[30]

Many of the first recruits had gained their wings by spring 1938 and they soon had a chance to show off their skills when the squadron took part in air displays for the first time. The first chance for the people of south Wales to see 'their' squadron was the Empire Air Day in May 1938. An annual showcase of the nation's air power, it featured displays held at RAF stations, with the first staged in 1934.[31]

Spectators were treated to thrilling displays of dive bombing, aerobatics, blind flying and more. The display for 1938, to be held on 28 May was chosen as the occasion for

Cardiff's annual air display too. It would be a red-letter day for 614 Squadron.

In the week before there were flypasts over the country by formations of aircraft, some over Cardiff. From 23 May, the Welsh city was treated to flypasts of Hawker Hinds, Bristol Blenheims, Fairey Battles and Avro Ansons. On Saturday, 28 May, the day itself, there was heavy traffic as a record crowd came to Pengam to view a display of military aeroplanes and equipment and see aerobatics and formation flying by Hawker Harts, a Hawker Fury fighter, Fairey Battle and Handley Page Heyford bombers.[32]

The celebrations resulted in a fine image of 614 aircraft over their home city. It showed a formation of three of their Hawker Hectors flying over Cathays Park in Cardiff, the administrative and cultural heart of the city. The starboard aircraft was flown by the future president of the Squadron Association, the then P/O Glen Pallot. A photograph was taken by Ron Grant, a photographer with 614 Squadron, whose day job was also as a photographer with the *South Wales Echo*, Cardiff's daily paper. The *Western Mail* of 30 May featured a picture showing a line-up of RAF aircraft on display, including Hawker Hectors of 614 Squadron at Pengam.[33]

The Empire Air Day display had shown thousands of Cardiff people that 614 had arrived as a fighting unit – with its own pilots, locally-trained and its own operational aircraft. The regulars who had set up the squadron prior to 1 June 1937 had done their job and it was time for most of them to move on. As Edgar Salmoni remembered, 'By the summer of 1938 the 818s were running the Squadron, with a little help from the remaining regulars. The pilots were earning their wings and the squadron was at last getting off the ground in the real sense of the word.'[34]

There were, however, some areas where help from the regulars was still needed to get the squadron operating effectively.

One early joiner, Ron Hann, recalled a regular sergeant who kept 614 well fed. Sgt Brain came to Pengam Moors in September 1938. The cook produced 'the finest meals I ever had the pleasure of whoofing down, in the whole six and a half years I served in the RAF.'

In particular, Ron remembered his Sunday lunches in the Airmen's Mess. As he lived in nearby Tremorfa, Ron could have gone home to eat while on duty, but Sgt Brain's cooking was too tempting. 'I grew so fond of his Sunday lunches that I told my mother that I wouldn't be home for lunch due to "extra duties" at the base. I'm glad she never discovered the truth of the matter, or I would have been flung out of the house for good.'

He remembered a cake that Sgt Brain made was a model of Pengam Moors, complete with hangars and aircraft, all shaped in icing-sugar, 'a real masterpiece of confectionery'.[35]

With the Hector delivered and newly-qualified pilots gingerly gaining experience on it, they now could move forward and train in earnest for the job they were tasked with – Army Co-operation. The job was one of military aviation's earliest – that of spotting for artillery and giving warning of enemy troop movements and positions. It also included aerial photography and supply-dropping. They were the eyes of the army on the ground and would live up to the squadron motto, 'Codaf i Geisio' – 'I Rise to Search or Seek.'

To begin to learn that role properly, they would have to move to the other side of the country.

A fundamental feature of AAF squadrons and all Territorial units was (and still is to an extent) the annual camp. While much of the year had training at weekends and on week nights, the summer camp lasting up to 15 days, was used to put into practice what had been learned across the year, so personnel would be trained and ready should war come.

The squadron's first summer camp was to RAF Hawkinge in Kent. A centre for No. 22 (Army Co-operation) Group, this was the perfect place for the new squadron to learn their trade.

Held between 2 and 17 July, it would be the biggest logistical challenge the squadron had faced and would involve moving much of the unit en masse 220 miles away to Kent.[36]

Over 50 men and all the squadron's aircraft and equipment went by air, train and road to Hawkinge. An early document from this important first step in the squadron's development is the Movement Order to go to Hawkinge. This was the plan for how the squadron would move across the country.

It detailed how an advance party would go by rail on 26 June, followed by Flt Lt WB Frederick who was to fly to Hawkinge on 27 June. A convoy of two vehicles was also to make its way to Kent via Odiham.

Travelling on 1 July, the main rail party would be led by Sgt Tucker with AC Jenkins and LACs Bell, Pittam, Sage, Brind, Lloyd, Lewis and Larmond. They would be transported to Cardiff General Station, from where they would travel to London Paddington and then to Folkestone. They would be collected from Folkestone Central Station by the lorries of the road party who would have already off-loaded equipment taken from Pengam at Hawkinge.

The Air Party would fly five Hectors, one Hind, one Hart and three Tutors. Crews were to stand by from 14.00 on 2 July to fly across country to Hawkinge individually, if weather conditions were good.

There was also detailed guidance on kit and clothing the men were expected to bring. The order explained, 'All ranks are advised to take only a minimum of kit to camp, since accommodation will be under canvas.' Officers, however, would have to spend more time packing. They needed 'service dress, including hats, gloves and greatcoats, also Mess kit or failing that, dinner jacket.'

Thought had been given to airmen's accommodation in Cardiff while they were away. 'Airmen will continue to draw lodging allowance during camp and are to make arrangements personally to retain their lodgings. Retaining fees should not exceed lodging allowance which amounts to *8/9d.* per week.'

The order was signed, Richard Cadman, Squadron Leader, Commanding, No. 614 (a) Squadron, AAF. 19 JUNE 1938.[37]

Before the outbreak of the Second World War in September 1939, 614's early years were punctuated by darkening political skies in Europe. The rise of Nazi Germany from 1933 had prompted much of the expansion of the RAF and the AAF, which saw 614 Squadron come about.

In March 1938, Nazi Germany annexed Austria, a union between the two German-speaking nations that had been forbidden by treaties at the end of the First World War. It was seen as another sign of German territorial ambitions in Europe.

This tension formed a backdrop to many a decision by early recruits to join 614. Richard Rhys recalled, 'Most of the chaps thought they ought to be in something. Hitler was making nasty noises at the time, so it was a case of getting some preparation in as a precaution.'[38]

In September 1938, the situation escalated with the crisis over the Czech Sudetenland and its disputed German population. Nazi Germany demanded that it should be able to annex this region and tension grew as German forces massed on the border.

In the UK, the prospect of war looked very real as German demands escalated. From the weekend of 24–25 September, the whole country began to prepare, with Air Raid Precaution stations opening. Gas masks were given out and Sunday, 25 September, was nicknamed 'Gas Mask Sunday'. On the morning of Monday, 26 September, war, to many, seemed imminent.[39]

In this climate of crisis, the UK's armed forces were put on alert. Many of the gleaming silver aircraft of the RAF were hurriedly painted camouflage, including Hectors of regular squadrons. Another Army Co-op squadron, No. 59, based at Old Sarum in Wiltshire, showed how the deepening crisis affected units which soon could find themselves at war.

On 26 September, their squadron diary explained, 'Emergency declared as a result of the international political situation, officers and airmen recalled from leave and orderly room staff and various personnel sent to mobilisation pools.'

As the situation worsened, they stepped up preparations. The station was blacked-out and their silver Hawker Hectors were painted in camouflage and dispersed across the aerodrome.[40]

War seemed certain. The *Western Mail* of 27 September printed a guide on how to build a 'garden trench shelter for six persons'. The same page told of 'Air Raid Precautions speeding up in Wales' with a total of nine miles of trenches being dug in Cardiff.[41]

The Czechoslovak government hoped that Britain and France would come to its aid, but the British Prime Minister Neville Chamberlain was intent on averting war. At Munich, on 29 September, Chamberlain agreed that the Sudetenland could be incorporated into the Reich while securing Hitler's recognition of the independence of the rest of the Czechoslovak state. The Prime Minister hoped this would mark the dawn of a new era of European stability.

He returned to the UK to wave his famous 'Piece of Paper' – an agreement signed by Hitler which he believed would guarantee 'Peace in our time'. A report from the *Manchester Guardian* captures the jubilation that greeted the agreement, 'He drove from Heston to Buckingham Palace, where the crowd clamoured for him, and within five minutes of his arrival he was standing on the balcony of the Palace with the King and Queen and Mrs Chamberlain.'[42] History, of course, would prove him to be wrong.

War would not come to Britain, or to Wales for now.

Following this temporary lifting of the clouds over Europe, there was little 614 could do.

The winter of 1938–39 was a particularly wet one in south Wales. Pengam was waterlogged for much of the time and 614 Squadron's activities were limited.[43]

By the time better weather returned and Pengam Moors dried out, 614 began to practice in earnest for the forthcoming Empire Air Day. The display would take on a more purposeful tone this year. The drumbeat of war was quickening and, that February, the first of two and a half million Anderson shelters began to be issued to households across the UK.

More than just a pleasant day out, the Air Day that year would show the capability of the RAF to friend and potential foe alike. Peter Farr recalled that the squadron was scheduled to do a nine-aircraft formation take off which they had March and April to perfect.

In those months, the situation in Europe deteriorated further. On 15 March, Nazi Germany invaded the remaining part of Czechoslovakia, breaking the Munich Agreement it had signed with Great Britain and France just six months before. Within a week, Hitler made new territorial demands – this time to return the Free City of Danzig to Germany. On 31 March, France and Britain pledged to guarantee Poland's security and independence.

Although Europe seemed to be teetering again on the brink of war, for Peter Farr, one March day in 1939, spring was in the air and, again, it was too fine a day for lectures. Playing truant once more, he called in at Pengam Moors. It was to be a fateful decision – one which would put him in great peril but would save the squadron from grievous loss.

He was met by the new adjutant, F/O Smith. Farr recalled that 'Smithy' had a job for him and was pounced upon as he entered the hangar. Smith said, 'Ah! Just the man. We want someone to fly to Old Sarum to collect some equipment for our Empire Air Day display.'

Farr completed his engine runs and checks and lined up for take-off against the easterly wind. Then, at full throttle, he took off at 90mph but suddenly the engine lost power, leaving the propeller ticking over. The sea wall surrounding Pengam Moors approached fast. He recalled, 'First, I sat there gawking at the propeller, and then I realized how serious the situation was. I could not possibly stop before piling into the six-feet-deep ditch, the six-feet-high sea wall, and after that the muddy and deep River Rhymney estuary.'

He had to do something. Farr closed the throttle and boost and saw there was no fuel pressure. But as he watched the gauge, the needle began to move up. 'I let it go for a few moments, and then eased the throttle open and to my relief the engine responded and the aircraft remained airborne.'

As disaster loomed, time slowed, although it all happened in about five seconds. 'Any longer,' he recalled, 'and I could not have remained airborne.' Having cleared the ditch, the sea wall and the river estuary, he decided to remain airborne but did not dare open the throttle any further. He climbed slowly and after ten miles had managed to reach 750 feet. Farr now felt safe enough to turn back towards the aerodrome. When he'd reached 1,500 feet, he opened the throttle to test the fuel pressure, which immediately fell to zero, leaving the propeller wind-milling. He closed the throttle again and the engine sprang to life. He landed safely, but could not get the engine to stop.

'Arriving at the dispersal, all hell broke loose, because they assumed that I had not tested the engine before taking off. I got out of the aircraft, and invited the Flight Sergeant to check it, and after quite a long run, she stopped again.' On investigation, mud was found in the filters and the fuel tank. Worse still, every aircraft was found to be in the same state. It transpired that the mud had come from a replacement bowser hose that had been left lying on open ground during the winter. Without his impromptu flight – and the skill he showed and the luck he had – the squadron might have lost almost all its aircraft and its pilots. He recalled, 'That evening we had been scheduled to do a nine-aircraft formation take off and practice flying for the forthcoming display – imagine the chaos if I had not played truant!'[44]

Thankfully, the groundcrew were able to drain fuel from all of the Hectors and flush out any grit to stop the problem happening again. Practice for the Empire Air Day could continue.

The celebrations began that year on 8 May with formations of Fairey Battle light bombers swooping across the skies of south Wales. The Battles and many of the aircraft taking part would be in drab camouflage – reflecting a force that was increasingly preparing for war.

On the Saturday of Empire Air Day itself, crowds again flocked to Pengam with over 15,000 spectators. A Short Sunderland flying boat from RAF Pembroke Dock opened and closed the display with a flypast. Nine of 614's Hawker Hectors and its Avro Tutors also featured, with the Hectors giving that formation take-off they had practised.

The display that year gave prominence to the now well-established squadron and was well reported. The *Western Mail* of 19 May previewed the display with a picture of the CO and seven pilots of 614 in front of one of their Hawker Hectors. The following day a superb photograph of squadron aircraft featured with nine Hectors lined up, flanked by Avro Tutors on either side. It came with a report which told how a 'record crowd of 15,000 witnessed the display at Cardiff Municipal Airport'. It said the nine Hawker Hectors of 614 Squadron 'made a big impression in formation and the splendid symmetry kept throughout showed the constant practice carried out.'

'Machines of the squadron also took part in demonstrations of picking up messages [with a hook trailed from a low-flying Hector] and aerobatics.' There was a demonstration too of the squadron's latest type, the Westland Lysander, which showed 'its remarkably slow flying qualities' and dropped parachute supplies with 'great accuracy'.[45]

As the squadron prepared for its annual camp at Hawkinge that summer, more Lysanders arrived.[46] A monoplane with an enclosed cockpit, the striking Lysander was a huge advance on the Hector. The aviation writer Bill Gunston summed up the unique aircraft. 'Utterly unlike anything else in the RAF, the "Lizzie" had the look of a dragonfly about it, because of its long and slender wings seemingly about to part company with the tubby fuselage hanging underneath. People who were useless at aircraft recognition used to divide up flying machines into two groups: aeroplanes and Lysanders.'

A key requirement of an army co-operation aircraft was the ability to operate from fields near frontline troops. Its innovative wing design meant its STOL (short take-off and landing) performance was outstanding.

The ungainly-looking two-seater had outstanding all-round vision for its roles of artillery spotting and reconnaissance. The high-set wings tapered inwards so they did not impede the view from the cockpit where the crew sat high, looking out from a large greenhouse-like canopy.

It was powered by a Bristol Mercury engine of up to 905hp, giving it a higher performance and enabling it to carry heavier loads than the Hector. It also packed a bigger punch and was armed with a fixed Browning machine-gun mounted in the top of each wheel housing. The observer had a Vickers K or a Lewis Mk III machine-gun.

So equipped with this new aircraft, 614 prepared for its annual camp, again at Hawkinge, while tensions in Europe continued to build.

On 10 July, Prime Minister Neville Chamberlain reaffirmed support for Poland and said Britain would intervene on its behalf if hostilities broke out between them and Germany. The deepening gloom was presaged in his visit to Cardiff in

June. In place of 'Peace in our time', he talked of Britain's powerful navy and air force growing 'at a rate beyond our expectation'.[47]

Not long before the arrival of the Lysander, Ray Merrett joined the squadron in early 1939 and remembered attending training 'every Monday night, Thursday night, Saturday afternoon and Sunday'. Their training continued and he went to the camp that summer. He recalled, 'war was on the horizon right up until the summer – and there were lots of arguments over politics.'[48]

At Hawkinge on the Kent coast in late summer, France would have been perfectly visible. The scene would have been peaceful and utterly unlike the furious carnage that would rage across it just nine months later. By the time they returned from Hawkinge, the political situation they argued about had deteriorated.

As the squadron returned, troubling news came from the Continent. After Germany's invasion of Czechoslovakia in March, Chamberlain hoped to strengthen ties with the Soviet Union and form an alliance between themselves and France to guard against further German aggression. The hopeful overtures to the Soviets were still being made when news came, on Tuesday 22 August, which rocked Britain and France.

Newspapers reported that a pact between Nazi Germany and the Soviet Union was about to be signed. This had seemed impossible as Hitler had made it clear his hatred of Communist Russia, controlled, as he saw it, by sub-human Slavs and Jews.

Headlines read, 'Russia and Germany: non-aggression treaty to be signed – Ribbentrop flying to Moscow tomorrow.'[49]

The next day it was confirmed, the two sworn enemies had forged an agreement which committed both countries not to attack each other and which sealed the fate of Europe. Britain's attempts to create an alliance with the Soviets had failed – the British and French strategy lay in tatters. Parliament was recalled the next day to urgently debate the next course of action.

By Thursday, 24 August, Parliament passed the Emergency Powers (Defence) Act, giving the government broad powers in order to conduct war effectively.[50] An act was also passed ordering the embodiment of the Reserve Forces.

On Friday 25 August, just two days after the Nazi-Soviet Pact, the Agreement of Mutual Assistance between the UK and Poland was signed, promising mutual military aid if either was attacked.[51] On the same day, 614 Squadron received the order to be embodied (merged into the RAF and brought to active service).

That week, as the news worsened, 614 Squadron was caught in that frenzied time as war approached. Ray Merrett recalled that life for him could have taken a very different turn. 'I was 18 and working in a barrister's office as an articled clerk in the hope that, in September, I was supposed to be going from Cardiff to the London office to become a barrister.'

But in August 1939 everything changed. 'On the Wednesday night I was sleeping in a feather bed with feather pillows at home and the following night I was down at Pengam sleeping on a bag of straw for a bed and another bag of straw for a pillow. I'd been called up. And that was it and I was, as I saw it, grown up overnight.'[52]

An early recruit, Hector MacDonald recalled they'd 'barely got home' from summer camp at RAF Hawkinge when, 'on the 26th August we were called up to do battle with the forces of evil'. This consisted, he remembered, 'of filling sandbags under the scrutiny of F/S Boreham, to protect the squadron HQ and guarding, with pick-axe handles, the many civilian aircraft that had been dispersed to Cardiff.'[53]

Like Ray Merrett, Cliff Haines was only 18 years old on the eve of war and remembers how bedding being handed out was somehow a marker for this move towards war and adulthood.

'General mobilisation was ordered in August 1939 and, like many others, I was called up. Being just under 18 years old my induction came as quite a shock when I was issued with three rough blankets, three planks of wood, two end pieces, and a paillasse mattress to be filled with straw, and then told to find a place in the drill-hall to sleep.'

He too recalled being on sentry duty with makeshift weapons. 'When we were ordered onto guard duties, we were issued with pick-axe handles and were marched out to stand guard on the sea wall and the entrances onto the airfield.' He remembers that initial period of mobilisation in August being taken up with kitting out and vaccinations, 'all of which,' he said, 'made us realize that this was for real'.[54]

As the patrols were stepped up at Pengam Moors in the last few days of August, Hitler stepped up his demands. On Wednesday 30 August, Hitler issued an ultimatum to Poland demanding that she give up the Polish Corridor (the strip of land allowing Poland access to the Baltic Sea) and the Free City of Danzig. Germany was on a collision course with Britain and France who were committed to helping Poland.

Two days later, on Friday 1 September, Germany invaded Poland. Although the UK was not yet at war, the evacuation of civilians from large cities began. Air Raid Precautions were introduced and blackout enforced from sunset to sunrise.

The next day, some normal life continued. Cardiff City played Notts County at Ninian Park before a crowd of 20,000; but it was to be their last football league match for seven years.[55] Chamberlain and the French Prime Minister Édouard Daladier issued an ultimatum to Germany, demanding it withdraw from Poland within twelve hours. On Sunday 3 September, at 11.15, Chamberlain announced on the radio that as Germany had failed to do this by 11.00, a state of war existed between the United Kingdom and Germany.

The British government announced the immediate closure of theatres, cinemas and music halls. Football matches and other events likely to be attended by large crowds were banned, as was the blowing of whistles or sounding of horns.

In keeping with Europe's dire situation, on the opening day of the war, 614 Squadron lost one of its precious aircraft. Hawker Hind L7238 nosed over in a field at Magor on the Gwent Levels after its engine cut out at 2,000 feet. As the squadron diary reported, 'The field was of a boggy nature, but was the only possible landing site in the vicinity.' The Hind was written off, but the pilot, F/O Smith, was unhurt.

With the outbreak of war, the squadron received word on 9 September that it was to leave Cardiff. On 2 October, they departed for Odiham, the Army Co-Operation aerodrome in Hampshire. Dai Park recalled it was a low-key affair; there was no civic send-off. 'We marched out of Pengam down Whitaker Road to board the troop train at the Roath Coal Sidings on Newport Road. The only people who were there to wish us good luck and wave us goodbye were families, friends, neighbours and the girls from the armaments factory on Pengam Bridge. No civic dignitaries for 614.' But among those family members who bid them farewell was Dai Park's father. He had been gassed in the First World War at Ypres, and struggled physically as a result. Dai suddenly saw him at his side. In spite of his frailty, he came and walked alongside his son as he marched to the railway station.[56] And with that, 614 Squadron went to war.

With little fanfare, the squadron moved to their new home. The squadron arriving at Odiham consisted of 24 officers and 136 other ranks (ORs). A further 156 ORs were left behind in Cardiff as they were 'insufficiently trained to have any practical use'. These would be sent to training camps,[57] and while some would return once their training was over, others would never serve with the squadron again. Alongside the Lysanders, two surviving Hectors, K8111 and K9740, were also flown to Odiham. The three original Avro Tutors still survived and one, K6103, also came with the squadron.[58]

Cliff Haines was sent to Hednesford to train as an Aircraft Fitter and would return to 614 Squadron. Hector MacDonald, too, left for training but his time with the squadron was now at an end. 'In November, most of the u/t W/Ops [under training Wireless Operators], were posted to No. 2 Electrical and Wireless School at Yatesbury and so ended my career with 614.'[59]

He recalled a short-lived meeting with another '818' at Yatesbury. Eddie Parrish (818132) would be transferred to 49 Squadron and Hector remembers that 'sadly he was killed over Germany within a few days of our reunion'.[60]

These changes, as 614 Squadron transformed from Auxiliaries to a regular unit, were swift. Another five pilots were sent on an Army Co-Operation course at Old Sarum, with a further five being sent to complete flying training at 5 FTS, RAF Sealand in north Wales.

On 10 October some more of them learned they would be leaving. 'It was decided to form No. 225 (Army Co-operation) Squadron from ['B' Flight] 614 Squadron,' the squadron diary reported. As a result, six Officers would leave 614. They included some who had joined in the very early days of the squadron, like Bill Drummond and Richard Rhys. As well as the officers, 56 ORs left 614 to form 225 Squadron. Five Lysanders were also transferred to the new unit, which would also be based at Odiham.

By the start of November, units of the RAF had been sent to France as part of the 'Air Component' of the British Expeditionary Force (BEF). Although not on the front line, 614 had contact with units which had been sent to the Continent, with Flt Lt Lacey and Sgt Morton ferrying aircraft to France.

When the squadron left Pengam, Sqn Ldr Cadman remained to become Station Comander RAF Cardiff (the new name for RAF Pengam Moors). He was replaced by Flt Lt LJ Stickley, who, in turn, was replaced by Sqn Ldr WR

Wills-Sandford as CO on 8 November. Some more men from the Auxiliary days in Cardiff left the squadron, with P/O Potter returning to Pengam that day. F/O Charles Cadman (Sqn Ldr Cadman's son) left to be attached to the School of Administration.

There was further contact with the Air Component in France when Sqn Ldr Wills-Sandford and P/O Pallot flew to France to return the next day. Meanwhile, as if to underline that there was in fact a war on, Flt Lt TK Lacey was fired upon while over London on a photographic sortie.[61]

The new year saw another change of leadership when, on 12 January, Sqn Ldr Wills-Sandford was posted to HQ No. 52 Wing RAF Component in France. In what would be an increasingly close relationship with the Army, Sqn Ldr AAN Malan (ex-Major, Royal Tank Regiment) assumed command of the squadron.

A reminder of the closeness of the co-operation between Britain and France came on 30 January with a visit from French Deputies (MPs) to Odiham. The links with the RAF in France continued during March when, intriguingly, given the squadron's future role in delivering chemical weapons, officers from the Air Component in France were attached to the unit for the gas spray course.

In April and the first few days of May, the postings, training and exercises continued. On 1 May, Flt Lt Rhys-Jones, one of the first intake of pilots at Pengam, was attached to Ringway on temporary duty. Five days later, 614's numbers grew with the posting of five pilot officers to the squadron from Old Sarum. On 9 May, another of the first intake, F/O Richard England, forced landed at West Down Artillery Range. There is no mention of injury to him but his aircraft was written off. However, those gentle comings and goings, and mishaps would soon be rudely interrupted.[62]

Following the outbreak of the Second World War in September 1939, its opening act, the German invasion of

Poland, was swiftly concluded. Although bound by treaty to defend her, Britain and France could do little to help and stalemate ensued. The Allies quickly settled into what the British called 'The Phoney War' and the French, 'La Drôle de Guerre' (the funny war). The Germans had a name for it too – 'Sitzkrieg'.

The events of April and May would see all that changed. In the early morning of 9 April 1940, the Germans launched their invasion of Denmark and Norway. But worse was to come. A month later, on 10 May, German troops and tanks crossed into Holland and Belgium. As the British and French rushed north into Belgium to meet the German attack, further south German forces invaded France through the dense forests of the Ardennes. The Allies thought the forest was impassable for armour and were taken completely by surprise. Suddenly the 'Phoney War' was phoney no more. As the world turned upside down, it seemed 614's world would as well. But what would they be called on to do?

Alongside the routine business recorded in the squadron diary, there was an 'additional note' for May. It read, 'There has been very little flying done since the German invasion of the Low Countries. Since then, the squadron operated entirely from a dispersal area at nearby Insham's Farm.'

Just a week before the invasion on 3 May, six aircraft of the squadron were ferried to Amiens in France with crews from the squadron. It had seemed a routine operation, much in keeping with the 'Phoney War' but, within just seven days, Amiens would be near the front line and by 20 May would be occupied by the Germans.

Through that feverish month, other RAF Army Co-Operation units flying in France were cut from the sky in their ungainly Lysanders – some of them supplied by 614. At Odiham, the squadron stood ready. Towards the end of the month, as the situation became ever-desperate across the English Channel, it seemed 614's time had come. The squadron diary recalled the alert, 'On the night of 24 May, 13 aircraft were got ready and equipped to drop supplies. The pilots and air gunners were standing by to take off at 4am on 25 May.'[63]

In the end, it was another Army Co-Operation squadron, indeed another AAF unit – just one digit away from the Welsh squadron, at that – No. 613 (City of Manchester), which was called. Based alongside 614 at Odiham, it was they who'd fly to Calais in a type even more outclassed than the Lysander which 614 had by then got rid of – the Hawker Hector. On 25 and 26 May, their Hectors attacked German troops surrounding Calais as supplies were dropped to the British by their Lysanders. The 170mph biplanes were completely outclassed by the Luftwaffe's modern fighters which dominated the air. It was to be the only time the outmoded aircraft saw combat and it was a miracle that they were not shot out of the sky.[64]

Ironically, one of the pilots involved in the operation, P/O 'Paddy' Barthropp had been flying in France with No. 4 Squadron, but was transferred to 613 Squadron on 3 May with another pilot who was now with 614 – Peter Le Cheminant. But when it was realised that, due to a typing error, Le Cheminant should have gone to 614, he joined the Welsh squadron. On that night their fates parted again as 613 Squadron took to the skies in their Hectors to do battle against the advancing Germans. Recalling the day he joined on 4 May, Peter remembered 614 Squadron was very proud of its Welsh roots and were 'Taffies to a man' and the Guernseyman would stay with the Welsh unit for the next two and a half years.

That month, news came that the squadron would move again. With war coming closer to southern England, where 614 was based – would they go towards the thick of the fight? On 7 June, as German forces swept across France and the last of the evacuated troops from Dunkirk licked their wounds, 614 Squadron began its next move. They would be going to Scotland.

CHAPTER 2

Scotland (1940–1942)

In June 1940, Britain gathered her strength to repel an expected invasion. Germany had defeated all the major armies of Western Europe. Why should Britain be any different? Around 300,000 soldiers had been rescued from Dunkirk but all of their equipment, save their rifles – their tanks, their heavy guns, their lorries – was left behind.

Alongside this threadbare army stood the RAF in defence of the islands. But while much of the Air Force's strength was concentrated further south, 614 Squadron was heading for Grangemouth, near Edinburgh. Most of 614's Auxiliary counterparts like Nos 601, 600 and 611 Squadrons were flying Spitfires and Hurricanes from airfields in the south-east of England in what was already being called the Battle of Britain.

No. 614 were given a job – not as glamorous as the others – which would prove to be a frustrating one. Their role would be to patrol the long coastline of eastern Scotland, watching for signs of the expected German invasion. If they came, 614 were to raise the alarm and try to stop them before they could land. It would be a time when the squadron trained to unleash the most horrific weapon in the country's arsenal, and when the first of its men were killed.

Although many expected the invasion to come across the

English Channel in south-east England, 614 was sent north out of fear of a lesser, but still real possibility – a strike launched from Norway.

Scotland was certainly vulnerable. When Germany invaded Norway in April 1940, the rapid capture of airfields was a graphic demonstration of their ability to seize vital targets using airborne troops. If the Germans could mount an operation like this, 300 miles from their bases in northern Germany, what was to stop them doing the same from their newly-won territory in Norway and flying to Scotland?[1]

The risk of just such a German invasion of Scotland was envisaged in a paper put to the Chiefs of Staff committee on 25 July 1940 by Admiral Sir Reginald Drax, who was responsible for protecting the sea lanes on the east coast of Britain.

In the wake of the German victory in Western Europe, his paper argued that the Germans could put 50,000 troops into northern Scotland and within a few days concentrate such forces around Wick to make the naval base at Scapa Flow untenable.[2]

But there were earlier signs that preparations for this actually might have been underway. The following signals were intercepted in May 1940. 'Warning message from 9th (Highland) Division re notice from Admiralty of concentration of flat-bottomed ships in NORWAY.'[3] This was from the *War Diary of 28th Infantry Bde*, 10 May 1940.

Another read, 'Troop carrying aircraft assembling STAVANGER. Possible attack Orkney and Shetland.' This from HQ of 26th Infantry Bde, 'Urgent message to 5th Btn Queen's Own Cameron Highlanders, Tain [the northernmost infantry unit on the British mainland], 17 May 1940.'[4]

Just how real this threat of invasion was felt to be was plain as the squadron travelled north. On 6 June, just as the squadron was arriving in Scotland, instructions were given on how Army Co-Operation units like 614 were to operate in the event of a landing. The order from No. 22 Group said that its squadrons were to stand by for immediate attack on the orders of Army formations. 'These instructions,' the order read, 'were drawn up to provide bombing of enemy transports and troop landings during the critical period of at least one hour between the initial report of invasion.'[5]

Arriving at Grangemouth, 614 was soon exercising in reconnaissance to sharpen skills dulled by weeks of fruitless waiting at Odiham. As if to emphasise this new threat, on 14 June, Flt/Lt Martin Edwards was sent on a sortie after reports of parachutists landing.

Their task was huge. On 17 June regular patrols began, covering the entire east coast of Scotland from Duncansby Head at the north-eastern tip of the mainland in the north to the border with England at Berwick-upon-Tweed in the south – 430 miles away.

The coast was dotted with beaches where troops could disembark and any number of fields where parachutists could land. It was 614's job to watch the seas and the skies along this vast extent of coast and warn of invasion when it came and do their best to stop them.

Five patrols were mounted, each at dusk and dawn. Three were from Grangemouth and two carried out by 'B' Flight, based 150 miles north at Inverness. The patrols covered the coast and five miles inland from Berwick-on-Tweed north to Montrose (approximately 140 miles); Montrose north to Kinnaird Head (85 miles); Inverness to Kinnaird Head (95 miles) and Inverness to Duncansby Head (120 miles).[6]

Detached flights went to Montrose and Evanton (which soon moved to Inverness). Peter Le Cheminant was based at Inverness and recalled the strangely unwarlike posting. 'We were billeted in the Caledonian Hotel in great style and flew from a short grass strip near the gasometer, which was what then passed for Inverness Airport.'

He recalled the coastline was 'virtually undefended'. In their Lysanders they were the first line of defence over those lonely stretches of seaboard. He wondered how effective they'd be if the Germans came. 'We were armed with 20lb anti-personnel bombs on wing racks, and after a while with ungainly Hispano-Suiza cannon attached to our undercarriage. What effect this armament would have had on any invasion barges was problematic.'[7]

The Lysanders also had twin Browning guns fitted to give them still more firepower. It wasn't much, but it was all they had.[8]

At first, the squadron lived under canvas but in July many moved to more solid accommodation. The HQ moved from a corner of the aerodrome to Claret Farm, just outside the airfield. The local school was requisitioned and housed 280 airmen. All on duty still had to sleep under canvas in the dispersal area, ready for action.

Defence of airfields was vital, but it wasn't clear who was responsible – the Army or the RAF? Later in the war the RAF Regiment would be formed, a dedicated corps of soldiers belonging to the Air Force and trained in airfield defence. But, in 1940, the RAF had either to rely on the Army or make do with what it had.

There was an 'unseemly struggle' between the services over who should defend airfields, which continued throughout 1940. Although the Army eventually were given the task, they didn't always embrace it with enthusiasm.[9]

The squadron had to make what preparations it could. From July, airmen were seconded to defence duties and many practised with rifles and Lewis guns on the range at nearby Turnhouse Airfield.[10] They also had makeshift weapons. Bill Hooper recalled some were more makeshift than others. 'Some of our pick handles were wrapped in barbed wire or had nails driven through the head.' He remembered they would later be given more effective weapons, albeit not very modern, 'How lethal we felt when we were finally issued with ancient American Ross rifles. How we cursed their weight, too.'[11]

LAC JAS Watson recalled other rudimentary weapons. 'Some had nail-studded clubs until the official issue of pikes made from bayonets welded to a length of steel tubing.'[12]

Later that month, the defences of Grangemouth were bolstered by a Heath Robinson-like improvised design. Simple to build, it was called the 'Armadillo'. Effectively a pill-box on wheels, it was a lorry with a wooden box protected by a layer of gravel or concrete behind the driver's cab, which was shielded by steel plates.

On 30 July, the squadron diary told how its third Armadillo was collected from the LMS railway works depot at Woverton. 'This is,' the writer noted approvingly, 'properly armoured and should prove extremely useful.'

The diary continued, 'The other two Armadillos from RAF sources are mounted on good chassis but have little armour for the driver and none for the engine. There are also four Armadillos from Army sources. All of these are on extremely old chassis, two fail to function and all are practically unarmoured. They are quite valueless and Scottish Command have been informed.'

Aside from the seven makeshift armoured cars and airmen armed with pick-axe handles, other, more conventional, defensive measures were built up. In mid-July, machine-gun posts were erected in the walled gardens of two neighbouring farms. Later, four 20mm Hispano cannon arrived for station defence.[13]

Despite disagreements between the Army and the RAF as to whose responsibility defence of airfields was, 614 had support from some Army commanders in preparing Grangemouth. The Inspector General of Fortifications, Home Forces, Maj Gen GBO Taylor, visited and advised on placement of road blocks.

The Army was praised for the help it was giving. 'C' Company, 10th Btn South Staffs Regt, had 'rendered extremely valuable service with the defences of the camp throughout the month,' the diary noted, adding, 'They have been responsible for the entire trenching system for defence, the creation of dug-outs, sand bag, petrol stores, machine-gun posts, etc.'

'Full Defence Practice' was held on 5 August at 19.00. All units on the station, including the Royal Engineers, Argylls, Light Anti-aircraft detachments and the South Staffs took part and six Hurricanes of No. 263 Squadron acted as attacking aircraft.

Three days later, the station was visited by the Inspector of Explosives, a Mr Busteed, and the diary noted that 'about 1,500,000 rounds of ammunition are held on this station'. Would that be enough if the Germans came?[14]

Among the uses for those rounds were machine-guns and cannon of the squadron's Westland Lysanders. Throughout the squadron's time at Grangemouth, it continued to fly the ungainly-looking type.

There was another weapon the squadron had to hand – and much more sinister – that could only be used by the Lysanders as a last resort. It was being readied as Britain faced the prospect of invasion – and a fight for its very existence. That weapon was poisonous gas.

No. 614 was one of the few RAF squadrons to be trained and equipped to spray chemical weapons. Its place in a little-known select group is a mark of how desperate those times were. Whether those terrible weapons would ever have been used will never be known. But, at that time, as the prospect of an enemy invasion loomed, the intent from the prime minister and military chiefs was there and the capability, with chosen squadrons like 614, was there too.

The use of chemicals as a weapon dates back thousands of years, but public outrage at their use in the First World War led to the Geneva Protocol of 1925, prohibiting their use in warfare. The Protocol didn't stop their development and production and many states which ratified the Protocol reserved the right to use them against non-signatories or in retaliation if gas was used against them.[15] Britain was one of them.

At the outbreak of war Britain and France said they would abide by the Protocol. Hitler, in a speech, replied, 'whoever fights with poison gas will be fought with poison gas. Whoever departs from the rules of humane warfare can only expect that we shall do the same.'[16]

However, much had changed between then and June 1940. As Britain faced a Nazi-occupied Europe across the Channel, there would be a volte-face on the use of chemical weapons which would have a profound effect on 614 Squadron over the next two years.

On 15 June 1940, less than two weeks after the end of the Dunkirk evacuation, the Chief of the Imperial General Staff, Sir John Dill, sent what has been called 'one of the most explosive memoranda of the war'. Secret for over 30 years, it was entitled, *The Use of Gas in Home Defence*, and it explained the military argument for using mustard gas against an invading German army.[17]

Dill observed that there were strong military arguments for using gas, 'Enemy forces crowded on the beaches, with the confusion inevitable on first landing, would present a splendid target. Gas spray by aircraft under such conditions would be likely to have a more widespread and wholesale effect than high explosives.' He did see risks, such as losing American sympathy or the Germans retaliating in kind. But he thought they were worth taking 'at a time when our national existence is at stake'.[18]

His proposal faced fierce opposition from the Cabinet but support came from Churchill. He said, 'Supposing lodgements [ie German beachheads] were effected on our coast, there could be no better points for application of mustard. In my view there would be no need to wait for the enemy to adopt such methods... but the question of actual employment must be settled by the Cabinet.'[19]

Moves were already being made to equip RAF units should the fateful decision come. This entry from the No. 22 Group war diary from 27 May 1940 identified the five squadrons chosen to deliver the weapons, 'Chemical Warfare. Instructions received from Air Ministry that certain Lysander Squadrons would be required to carry out gas spray tasks if this form of warfare came into operation. The squadrons selected were Nos 2, 4, 26, 225 and 614 Squadrons.'[20]

If the Germans had come, the invasion would have been met by them flying their ungainly Lysanders loaded with spray tanks holding mustard gas. These tanks were named enigmatically SCI or 'Smoke Curtain Installation'. Its name reflected that they could lay either a smokescreen/curtain or poisonous gas.[21]

That 22 Group diary entry from 27 May 1940 set out just how many SCIs 614 and the other four units would receive, 'Each of these squadrons received 47 250lbs charged SCIs, ie sufficient for two squadron sortie loads at 12 I.E. [Installation Equipped] aircraft per squadron.'[22]

If gas was used, the small stocks of SCIs meant that the RAF had to plan to mount the maximum number of sorties in a single day. Having made its bombing run over the beach-head and released its gas, it was calculated that each aircraft 'should be able to return the empty tanks to a landing ground near the charging station, and pick up full ones without delay.'[23]

In late June 1940, training for those five squadrons increased. On 29 June, the 22 Group diary told how one officer pilot and an armament SNCO of each squadron attended a two-day course at RAF Halton on loading and decontamination of SCIs. It added, 'During the months of September, October and November a considerable amount of practice spray exercise was carried out by squadrons in conjunction with the Army.'[24]

LAC Watson vividly recalled the gas spray exercises,

often involving low-level attacks across the airfield by 614's Lysanders and, later, Blenheims. 'Obviously the gas was simulated but nevertheless quite horrific to watch, as it combined with a morale-breaking siren… which shrieked like all the demons below.'[25]

An RAF armourer who served with No. 4 Squadron (who, like 614, also flew Lysanders in the Army co-operation role) recalled working with poison gas. Hugh Martin remembered handling the deadly gas canisters and practising arming Lysanders with them. He recalled this did not always go smoothly. Once, an electrician working on the bomb release mechanism on a Lysander accidentally wired the left bomb release to the right bomb rack, and vice versa. While testing the aircraft on the ground, an attempt to release the empty rack on one side caused a poison gas canister to drop off the other wing and start leaking, much to the alarm of those present. There were several incidents where gas was spilt, though no-one was hurt.[26]

Of this deadly weapon, Ron Hann said, 'an overriding memory of Grangemouth was handling cylinders of mustard gas with which the Lysanders were constantly armed'. Neighbouring civilians were blissfully ignorant of the dangerous substance kept there. 'I often wonder what the townsfolk of Grangemouth would have said if they knew about the gas storage pits at the nearby airfield.'[27]

So, armed with gas, cannon or bombs, 614's Lysanders were Scotland's first defence against attack. A future CO of 614 who first joined the squadron as a junior pilot at Grangemouth was Charles Bonner. He spoke warmly of the 'Lizzie'. 'As an aircraft I truly loved the Lysander. A joy to fly – stable, quiet, well-behaved in all flying conditions and docile, it only growled, as it were, when the slots shot in and out when you approached the stall, at around 55mph.'[28]

It was perfectly designed for the Army Co-Operation role Peter Le Cheminant remembered. 'Its high wing, fitted with slots and flaps, gave it its remarkable short field performance. It was slow and extremely stable so that one could easily fly it with one's knees, which was just as well as a Lysander pilot was busy indeed.'

A Morse buzzer on the cockpit's starboard side allowed the pilot to communicate by wireless transmitter. Pilots were supposed to be able to send and receive at 15 words a minute, a feat which, he admitted, was beyond him.

The flyer had a message pad strapped to his right knee on which he wrote the details of what he saw, if doing 'Tac/R' (tactical reconnaissance), or plotted the fall of shot if spotting for guns in what was known as an 'Arty/R' (artillery reconnaissance).

The pilot also had to look after aerial photography – either oblique or vertical. 'All of this,' he said, 'was fine in peacetime conditions but was not viable in the face of fighters and anti-aircraft fire.'

Earlier in the month, 614 was to lose more Lysanders, but with far more serious consequences. Le Cheminant recalled a 'fairly indulgent view' was taken towards accidents in 1940; they were 'if not inevitable, were certainly very likely and that, if you survived, you would have learnt something and thereby become a better pilot.'

One accident on 10 August was to have dire consequences. He'd been detached to Montrose and, after having flown the southerly dawn patrol that morning, was detailed to practise evasion tactics with another aircraft in the afternoon.

The other aircraft was piloted by F/O (Acting Flt Lt) Norman Stuart Merrett. One of the original intake of pilots, he joined 614 Squadron at Pengam Moors in 1938. The tactics they practised could, Le Cheminant said, easily degenerate into a dogfight, 'and that is what happened on this occasion,' he recalled. He remembered, 'Something went disastrously wrong and we were suddenly on a head-on collision course only yards apart. We collided starboard wing to starboard wing with Merrett's wing being slightly lower and hitting my struts.'[29]

Merrett's aircraft lost practically the whole of its starboard wing and spun to the ground just outside Montrose. Both he and his passenger, F/O James Ferguson Harper, the medical officer at Montrose, were killed.[30] They were the first of 614 to be killed in peace or in war.

Le Cheminant's Lysander somehow remained in the air. He managed to land safely and, as was customary then, he was immediately sent up again to ensure his nerve hadn't gone.[31] Neither Le Cheminant nor his passenger, Sgt PJ Ellwood, were hurt.[32]

F/O Harper was from Renfrew and is buried at Arkleston cemetery in the town.[33] Norman Merrett's final journey, too, was to his home. He is buried in his home village of Michaelston-le-Pit, just outside Cardiff. Heartbroken by the news of Norman's death, his father Herbert Merrett inspired villagers to join the nationwide Spitfire appeal – where whole communities would raise money to pay for a fighter. Within ten days of his son's death, the hundred inhabitants of the village raised enough to pay for a Spitfire which was sent to Lord Beaverbrook, Minister for Aircraft Production.[34]

With the cheque was a letter in which Sir Herbert wrote, 'I cannot provide you with another gallant son. The one who has gone was my only son. But I want you to accept from the village of Michaelston-le-Pit the enclosed cheque for £5,000 to purchase a Spitfire.'[35] He also had a stained-glass window to honour his son erected in the local church where Flt Lt Merrett is buried.

But the squadron had to carry on. The threat of invasion loomed still and the tempo of preparations increased.

As dusk fell on 12 August, there was a full defence practice involving RAF personnel and attached Army forces. Hurricanes of No. 263 Squadron and one platoon of the South Staffs acted as the enemy.

Towards the end of the month their defences were boosted with the arrival of four 20mm Hispano guns for airfield defence. One officer and 64 airmen were also posted to the squadron for defence duties.[36]

Early September saw the German attacks intensify and the threat take on a new, more menacing form. It was a change which would mean relief for some but would mean imminent danger for others. Almost daily, signals came in which all seemed to point to one thing – that the expected invasion was near.

In reprisal for German bombs dropped on London on the night of 24–25 August, a raid on Berlin was launched the following night. Hermann Göring had insisted that such an attack was impossible because of the city's formidable air defence network. The raid so infuriated Hitler that his previous order banning attacks on London was rescinded.[37]

The results of this retaliation would soon come but, before then, on 3 September, four German spies were landed by boat on the southern coast of Britain. All were caught within hours and it was discovered their task was to report on coastal defences, military strengths and movements.

Following interrogation, the four spies said they were an advance guard for the invasion, which they said could be expected at any moment.

On 5 September, RAF reconnaissance missions reported large increases in the number of barges at French Channel ports, expected to be used to ferry troops. The next day, German bombing raids on port installations along the south coast of Britain led to the issuing of the 'Yellow' invasion alarm: 'Probable attack within three days.' Hitler now ordered German bomber raids on London.[38]

Different strands of intelligence now suggested strongly that an invasion was imminent; a 'striking increase' in the number of barges and other vessels moving to French Channel ports; the movement of two German Air Force groups to the

Low Countries and of dive-bombers to the Straits of Dover. The moon and tides were also favourable for a landing between the 8th and 10th.[39]

On Saturday, 7 September 1940, British Intelligence tried to work out why large numbers of barges were moving to French Channel ports and why German Army leave had been cancelled on the following day.

That same afternoon, 348 German bombers and more than 600 fighters set off from their bases in northern France for England. The bombers would drop 337 tons of bombs on the capital, killing 448 Londoners.[40]

Air Chief Marshal Hugh Dowding, Air Officer Commanding Fighter Command, later recognized the relief it gave the RAF from the pounding they had faced, calling it 'a supernatural intervention at that particular time and that 7 September was really the crucial day'.[41]

However, as Fighter Command could take a breath and regroup, other equally dire outcomes loomed. Was this the sign that the invasion was coming? It seemed that it might be imminent and this deduction was given to the Chiefs of Staff at 17.30. They concluded that, 'German preparations for invasion were so advanced that it could be attempted at any time.' The code word 'Cromwell' signifying 'invasion imminent' was issued.[42]

Scottish Command sent 'Cromwell' to its formations, including 614 Squadron. The 5th Division received the signal at 21.05[43] and cascaded the order to its brigades and individual units, most of which 'stood to' through the night. At 22.10 the 153rd Brigade in Aberdeenshire was ordered to five minutes' readiness to move. Church bells were rung in Stonehaven, 50 miles north of Dundee, following suspected sightings of German troops.

HQ 46th Division in Stirling alerted its 137th Infantry Brigade to reports of enemy landings by parachute and small boats at 01.15.[44] By 01.30 all units of 153rd Brigade were ready to move, and roadblocks and defence posts were manned.[45] The Royal Engineers in Edinburgh prepared to blow up Burntisland dock, across the Firth of Forth from Grangemouth.[46]

At 22.00 a flight of 614's Lysanders was bombed up and all crews from the three flights were standing by. Armed with 20lb-anti-personnel bombs and Hispano-Suiza cannon, its aircraft were ready. At 02.30, the diary told how 'most of the church bells in the district were rung'.

As dawn broke, the alert continued. The crews were poised to fly at minutes' notice. Eyes nervously scanned beaches and horizons – waited for signs that they were coming. But they did not come. At 11.25, 'Cromwell' was cancelled by Scottish Command. But for 614, a state of high alert would continue and they remained at readiness before being stood down at the end of September.[47]

Peter Le Cheminant recalled, 'Our readiness was heightened for 21 nights running. "B" Flight pilots slept in the dispersal hut and stood-to at dawn. It was a relief when winter came.'[48]

Although the immediate threat of invasion seemed to have lifted, the risk had not gone away and preparations continued. By late September, Scottish Command forecast that, 'Weather conditions are likely to hamper severely any enemy attempt to invade Scotland this winter. In spite of this, reasonable precautions against an attempt must be maintained.'[49]

There was other grim news for the squadron. On 26 September, dispatch rider AC1 William James Dowds died following a head-on collision with a car while travelling from the dispersal area of the airfield.[50]

Into the autumn, preparations went on. On 3 October, the Army assisted with the training of airfield defence personnel when 'C' Company of the Duke of Wellington's Regiment agreed to lend a sergeant instructor. The diary noted that there was a 'very great need for good drill instructors for ground defence personnel, many of who are very raw recruits'.[51]

Some of the squadron clearly needed sharpening. Around this time, SAC Donald Mackenzie recalled a visit from the Luftwaffe while airmen were on the parade square. 'A Dornier bomber suddenly appeared in the circuit going the wrong way round. For a while everybody just looked at it and nothing happened. Eventually, a Bofors gun opened up. I can still remember the tracer dropping hundreds of yards short. I don't think the Dornier pilot even opened up his throttles, it looked as if he had just eased back on the stick and disappeared up into the cloud.' He recalled that only then did the air-raid siren go off.[52]

Later in the month the CO, W/Cdr Eayrs, attended a meeting in Prestwick 'in connection with the "Banquet" scheme'. 'Operation Banquet' was a last-ditch plan to use every available aircraft against a German invasion.

While 614 Squadron patrolled the coast of Scotland, watching for signs of invasion, hundreds of miles south, much of the RAF was still engaged in the Battle of Britain. No. 614 would spend 1940 on the sidelines of the world's first battle conducted entirely in the air, and it was a time of frustration for many in the squadron.

One way the squadron coud help was to volunteer for air-sea rescue duties over the English Channel. From September 1940, pilots from 614 flew from Tangmere in Sussex to search for downed fighter pilots. The squadron's Lysanders could drop dinghies, survival equipment and smoke markers from hard-points normally used for bombs. The fantastic all-round visibility of the aircraft made them ideally suited to the task.

P/Os Lenny and Brown were the first to be chosen – going to Tangmere on 20 September. On 3 October, P/Os Forsythe and Laird left to take their place, and two weeks later P/Os Johnson and Mackenzie replaced them. Peter Le Cheminant was one of those who volunteered in the first three weeks of November. He found it an unsatisfying time. 'It was frustrating in the extreme to be with the Hurricane pilots, but not be one of them. I remember in particular envying the pilots of 43 Squadron, "The Fighting Cocks". Their morale and exuberance were at a high level and their expectation of daily action gave their lives an edge which was altogether missing in a Lysander squadron.'

He added, 'I was scrambled on a number of occasions, mostly to search for downed Hurricane pilots and once for the crew of a Hampden. The information given to guide the searches was very vague and could cover a large area. Sadly, but not surprisingly, I made no sightings.'[53]

One 614 pilot, at least, was able to contribute more. F/O Brian MacNamara joined 614 at Odiham, on 8 October 1939. He volunteered to serve in Fighter Command and on 21 August was posted to No. 7 OTU at Hawarden in north Wales to train on Spitfires. After a brief conversion he was posted to another AAF unit, No. 603 (City of Edinburgh) Squadron at Hornchurch, on 31 August.

He was very successful with many kills the following month. He shared in the destruction of a He 111 bomber on the 7th, damaged a He 111 on the 11th, destroyed a Me109 fighter on the 14th, damaged a Do17 bomber on the 15th, and damaged a Me109 on the 27th. Two days later he helped down a Do17 bomber.

He is one of a very few RAF pilots to have shot down an Italian aircraft during the battle. On 23 November a fighter sweep of 29 CR.42 Falco fighters over the English Channel was attacked by twelve Spitfires from 603 Squadron, who shot down seven of the biplanes, with one falling to MacNamara.[54] He would stay with 603 until 1 February 1941 when he returned to 614 Squadron.

Others wanted to play their part. On 19 October there was a request for volunteers to fly as air gunner in a Defiant squadron. (The Boulton-Paul Defiant was a single-engined fighter with a turret.) P/O Gillespie stepped forward and was posted on 1 November.

Meanwhile, at Grangemouth, things were becoming more organised – not least in maintaining morale with better accommodation and entertainment. Work on improving living quarters continued in fits and starts. There was great difficulty in getting building materials and huts being built for kitchens, offices and accommodation were still incomplete.

By the autumn the squadron was being well entertained, with many events held in Grangemouth town hall and some big names were billed. On 3 November, a concert was given by the famous Cardiff-born Ivor Novello. The show by the composer of such wartime songs as 'Keep the Home Fires Burning' played to a packed house of airmen, soldiers and sailors stationed at Grangemouth. A week later, Alastair Sim, Leslie Banks, Gillian Lind and company presented *A Cottage to Let* in the town hall.

The last show of the year was something of a change from the usual fare. On Boxing Day, an 'all-in' wrestling match was held in the town hall. Two professional and three amateur bouts took place. The professionals included the Scottish champion and airman AC Ricardo, who had won an Olympic medal.

The new year began with heavy snow and bitter cold. Flying was difficult as all aircraft were iced up. In desperation, a high-pressure hose was tried to clear the ice. Hessian covers were being made for all aircraft in a bid to keep the ice at bay. There was also word that the squadron, having been at Grangemouth for barely six months, was to be on the move again. They learned it was to be to Macmerry, an airfield the other side of Edinburgh. The anti-aircraft defences at Grangemouth were clearly getting sharper, with a report on 24 January of a Ju88 over the aerodrome which was believed to have been shot down in the sea at Dunbar by their Bofors guns.[55]

Charles Bonner recalled that the squadron was training hard throughout the early months of 1941, working on bombing, artillery shoots, photography, flying in formation at night and cross-country. He was attached to 'B' Flight under the command of Flt Lt Richard 'Dickie' England (whose family was well known in Cardiff as potato importers), an early joiner from 1938. 'A' Flight was commanded by another from Pengam, Flt Lt Phillip 'Johnnie' Lysaght, of Caerleon, near Newport.

He remembered practising high-level dive-bombing from 10,000 feet at a target in the Firth of Forth. 'I aimed the nose of the old "Lizzie" at the target, and as the speed built up the noise in the cockpit, from the engine and air-speed, reached terrible levels. It was most frightening! However, the bombs were dropped and I gradually pulled out and returned to base, only to find that both wings had been pushed back, along with their supporting struts and that some of the fabric had been stripped off. "Silly Boy," said Dickie England, "You were supposed to make a tactical dive using evasive action, gradually lose height, then at 3,000 feet dive-bomb the target!" The aircraft never flew again and I certainly knew the meaning of terminal velocity after that.'[56]

As the month ended, a signal from the Group came through 'advising move to Macmerry within six days'. The logistics of moving the squadron's aircraft, men and the equipment they had gathered in the last year would be formidable. The distance, however, would not be. They were only moving 41 miles east.[57]

Their new home, Macmerry, was not known by that name when 614 moved in. Then it was called Tranent, the name of a nearby village. The airfield was located between this settlement and Penston and, confusingly, during the First World War it was known as Penston.

Whichever name is used, there are echoes in its history of that of Pengam Moors. The grass airstrip was home to

Edinburgh Flying Club from the late 1920s and passenger flights began from 1932. The airfield was to have been the site of Edinburgh's airport before war broke out but the Air Ministry took it over in July 1940, deciding to rename Macmerry as Tranent. However, the name change never appeared on any but a handful of official documents and the name Macmerry stuck.

To some of 614's airmen from Cardiff, the look of the area reminded them of home, too. Macmerry's site on the grassy banks of the River Esk were similar, recalled '818' Ron Hann, 'to the tide-fields at Pengam Moors, long marshy grass, mud banks and usually breezy at the mouth of the river'.[58]

When 614 arrived at Macmerry, in March, there was little accommodation for aircraft or men. There was a small hangar and a flying control building and, while the airfield was grass, the perimeter track was tarmac.[59]

AC2 John Baxter recalled, as well as the hangar there were dispersal points, which were concealed like bunkers at points around the perimeter of the field.

He said, 'There was no accommodation on the airfield so airmen were billeted in the town.' On the upside, this meant that many of 614 were looked after by local residents. Baxter remembers staying at Macmerry Primary School while the camp was being built.

He recalled, 'Trucks took us to Tranent for our meals. After a while, I moved to Church Street School in Tranent and later to Sinclair Hall near the back of the Crown Hotel. Our meals were prepared and served in Blair Hall by the local residents. The food was very good and the townsfolk made us very welcome. Several airmen also billeted with families in Tranent.'

They were fortunate to have a coal mine nearby with baths and showers. 'We could shower once a week, sharing facilities with the miners at the Fleets Colliery. Every day we would be picked up in Tranent and taken to Macmerry.'[60]

Fitter LAC Watson also recalled being housed in a school in Tranent, from where they would be marched to Macmerry Airfield. 'The officers,' he remembered, 'were billeted at St Germain's House'. He, too, was thankful for the hot water at Fleets Colliery and 'the relative luxury of the pithead baths. These were as you can imagine, very much appreciated!'[61]

Some were not so thankful. Bill Hooper had 'bitter memories' of this time, recalling the morning and evening marches from Tranent to Macmerry and back 'carrying our weighty and ancient American Ross rifles'.[62]

John Richards, from Llanharan, near Pontyclun, was an experienced WOP/AG and was delighted to be posted to the squadron following two operational tours. 'As a native of Glamorgan, I was thrilled to be informed that I was to join No. 614 Squadron.' He was soon welcomed at Macmerry as warmly as the rest of the squadron by local people. 'I was billeted in Tranent with a small family, by the name of Muirhead. The man of the house was a miner, and he, and his wife and daughter treated me as part of the family.'[63]

Baxter also recalled with gratitude the hospitality of the locals. 'After our day's work we would be taken back to Tranent where our recreation included the "CUDS" [the local cinema] and perhaps dancing in the town hall.' All this hospitality led to Baxter, like many 614 airmen, meeting his future partner in the village. He recalled, 'I was fortunate to meet my future wife and was made very welcome by her family who lived in John Street near the primary school.'[64]

As well as Baxter many other 614 airmen met their wives at Tranent or the surrounding area. Dai Park, an '818' from Cardiff, recalled that when the RAF moved in at Macmerry, a number of women from Tranent formed a section of the Women's Voluntary Service (WVS). 'Twice-daily, in all weathers, they would arrive at the airfield and hand out cups of tea and cakes to the lads and show a smiling face to many who were far from home.' They also ran a canteen in the village at the Miners' Institute. 'This,' he said, 'became a centre of relaxation to members of the squadron, a haven where you could do the things you couldn't do on the camp.'

Another who found his partner amidst this kindness was 614 airman, Jimmy Redfern. He married a local girl, Megan Dobson, who was one of three sisters who provided the WVS service at the airfield. Megan, along with siblings May and Nettie, also helped out with the service on the camp and at the Miners' Institute canteen.[65] William Cowell joined 614 Squadron while they were at Odiham and, while serving at Macmerry, he too met his wife who was from nearby Wallyford.

Some of 614 had already met their wives in Grangemouth. Bill Hooper met his wife at a social function there. 'At the Baptist church door, this was the girl I would eventually marry,' he recalled many years later 'and three fine children and 36 years of a happy, stable married life were the results.'

He said, 'The relationship with the people was fantastic and the crop of Scottish wives who came south at the end of the war was ample proof of this. The friendship and generosity of the Scots was heart-warming.'[66]

For many months after they arrived, Macmerry Airfield was a work in progress. By the summer, accommodation on the airfield had been built for around 1,000 personnel and the whole of 614 Squadron moved into Macmerry. Work continued on the remainder of the airfield's infrastructure, including six blast shelters each capable of holding two twin-engined aircraft. Each of these had a smaller shelter built into them to hold 25 men. Eight hangars were added, as well as technical buildings.[67] Between the blast walls of the bomber dispersal pens the airmen dug vegetable patches.

The squadron continued with other invasion preparations and exercises. On 8 August, a 2nd Lt Hallam was attached to the unit to train ground personnel in station defence. Two days later a Scottish Civil Defence exercise was held involving mock attacks on Dundee, Cupar and St Andrews.

There were efforts, too, to innovate and expand the squadron's effectiveness. On 23 August, F/Os McClymont, Clark and P/O Bonner finished a blind approach flying course at Horsham St Faith in Norfolk. This would enable them to land in bad weather and at night. By December, a new runway was completed, which the CO, Wg Cdr Skelton, tested out.[68]

Despite the efforts to keep the men sharp, frustration, and perhaps boredom, began to creep in. Peter Le Cheminant remembered from when they moved to Macmerry until they converted to Blenheims in the following July. 'We trained incessantly, though to what end was not entirely clear to us at the time.'

This is well remembered by Bill Hooper. He recalled that the endless training became too much for two 614 pilots. 'In sheer frustration after a particularly boring exercise over the Lothians, they dived down Princes Street in Edinburgh, below the level of the castle with their sirens screaming and pulled up over Waverley Station.'

Something which must have leavened the boredom, however, was the conversion to a new aircraft.

The experience of using the Lysander in the Army Co-Operation role in France had shown it was very vulnerable to enemy aircraft, and Army Co-Operation squadrons were re-equipped with faster, better defended aircraft.

On 12 July, the squadron was told they'd soon be getting the twin-engined Blenheim MkIV.

When it first flew in June 1935, the Blenheim could reach 307mph, 80mph faster than the fastest fighters in the RAF. By 1941, any such speed advantage had long gone, but it was still a leap in capability from the Lysander.

Its Bristol Mercury XV engines rated at 920hp, gave it a top speed of 266mph. The bomb bay could carry 1,000lbs of bombs with a further 320lbs externally. A retractable turret had two belt-fed .303 Browning machine-guns and a further one fixed firing ahead.

The Mark IV was known as the 'Long Nose' as the nose of the Mk I (the 'Short Nose') was extended by three feet. The top of the nose was lowered on the left side to improve the pilot's view, giving a strange asymmetric shape. The extra three feet meant it was less cramped but it was still not spacious. The navigator (often referred to as the 'observer') sat on the right side of the nose but with barely adequate head room, facing to the left to plot courses on a chart table. He had to lie forward to use the bombsight.

Both got into the cockpit through a hatch behind them on the upper fuselage. Just behind this, and in front of the turret, was another hatch for the wireless operator/air gunner (WOP/AG).[69]

That July and August saw intensive training on the light bomber, and by the end of August conversion was all but completed. From then on, as Le Cheminant recalled, 'the bulk of our flying was on Blenheims'.[70]

Now the challenges of flying this new, more complex aircraft began to emerge. In September, night-flying training was carried out. The results were unnerving. It seemed Macmerry wasn't ideally suited to the bigger aircraft. The diary noted tartly, 'The aerodrome is rather small for Blenheims at night.'

There followed a series of crashes and narrow escapes involving the Blenheim as 614 crews became used to the new aircraft.

As with the Lysander in the summer of 1940, Peter Le Cheminant was to have two incidents, the second more serious than the first. The first saw a lucky escape in November 1941 while taking off. His starboard tyre burst and his propeller hit the ground. He was forced to collapse his undercarriage and 'slide ignominiously along the ground on a full bomb load, hoping against hope that all pins were securely in place, and would remain so'.

On 7 January 1942, he and his crew departed Macmerry for Andover with four other Blenheims and crews. Three days later, as they returned, he became disorientated in a snowstorm and, in a desperate struggle to save his aircraft, thought the end had come for them. Running into heavy snowfall near York with the cloud base down to 300 feet, he resolved to turn back.

He recalled making the mistake of taking his eyes off the instruments and sticking his head in the canopy blister to his left to make sure that all was clear as he turned to port. 'When I looked back at my instruments my gyro-horizon had toppled, my altitude was about 1,000 feet and my airspeed was down to 100mph.'

As he had gone into the turn at 250 feet and 150mph, 'it was quite evident,' he recalled, 'that I had established a gentle climbing turn and gone onto my back in the process. Whether I was still on my back or had rolled through into the upright, or was betwixt and between, I had no immediate means of knowing.' He knew he was in trouble. His life and those of his crew were in great peril.

'There have been a number of times in my flying life,' he remembered, 'when I knew I had little chance of surviving, but on this occasion I thought I had none whatsoever. I remember a stillness as the aircraft hung there and a feeling of calm and resignation. I thought I was about to die and had no feeling of panic or even fear.'

Something, instinct perhaps, made him keep going. He centralised the controls and waited to see what would happen. 'The airspeed started to build up, slowly at first and then quite rapidly as the altimeter unwound. It was evident that we were in a dive, so I cut the throttles and started to pull back on the control column.'

His gamble worked. 'We broke cloud at 300 feet heading straight for the top of a haystack. It looked huge and vertically below us but I think our angle must have been less, possibly somewhere between 60 and 45 degrees, otherwise I could never have pulled her through. Heaven knows what "G" she withstood, but my navigator's metal seat in the nose just buckled under him.'

On landing he recalled, 'I put it to my crew that they would never wish to fly with me again but they, in their ignorance, maintained that I had done such a marvellous job that they would never wish to fly with any other pilot.' Le Cheminant had landed in a field, so small that the aircraft had to be dismantled and recovered by road.[71]

The fierce snow that day claimed another victim as another Blenheim of that flight crashed when P/O Winter force-landed at Acaster Malvis near York.[72]

Aircraft Fitter Tom Hinchey was the sole crewman with P/O Winter that day. 'Three planes were in the flight, and we were relying on the lead aircraft for direction, everything was alright until we reached Yorkshire where we lost the others and ourselves in a blizzard.

'The pilot made about three attempts to land, but each time we almost hit power lines and telegraph poles as we came down through the clouds. We were running out of fuel and things were rather desperate when we made the decision to attempt another landing. This time we found luck and cleared the obstacles, finally crashing onto the snow-covered field, ploughing through a hedge and turning on our side as we came to a halt... no marks for speed of exit, suffice to say it was quick in case the aircraft blew up.'[73]

After these narrow escapes 614 was lucky to have three crews survive crashes. But that luck was not to hold.

On 13 February the Blenheim Z6033, flown by P/O Mummery, was taking off from Inverness when its port engine failed and it crashed in the shallow water of the Moray Firth. P/O Mummery and Sgt Field escaped, 'but,' the diary related, 'owing to the current, observer Sgt Nightingale was unable to cling to the aircraft and was drowned.' The other two were unharmed. The next day Flt Lt Munro went to Inverness and collected the survivors.[74]

There were many exercises preparing to repel an invasion, even though the prospect of one was fading. One of them saw 614's Blenheims detailed to 'bomb' Dundee as an enemy force, complete with black crosses stuck over the roundels and bags of soot as bombs. WOP/AG Paddy Stapleton recalled sitting in the turret as they flew at near rooftop height over the city and had a vivid memory of a woman, who was laying the table in a top-floor room, drop her tablecloth as she saw them fly past.

The frustrations from these seemingly endless exercises and the feeling they were on the sidelines of the 'real' war meant the squadron appeared to be stagnating.

By September 1941, Peter Le Cheminant felt despondent. 'I still had the feeling that we were never going to get into action.' He added, 'I think all the pilots felt the same and we had become rather slack, not really believing in what we were being required to do.' He wondered if this had been noticed higher up when a new CO was appointed.

He was Wg Cdr Dick Skelton. A seconded Army officer, Peter thought he 'had been briefed to smarten us up'. Skelton set about the task with apparent enthusiasm, instituting early morning parades, initially in full webbing with water bottle, pistols and gas masks, the men with rifles. This wasn't universally popular and Le Cheminant thought he knew why.

'The squadron had been diversified somewhat by the addition of four seconded Army officers and a few VRs [volunteer reserves, or wartime recruits], but the old pre-war hands, the senior Flight Lieutenants [such as Lysaght and England] were Auxiliaries to a man, all having joined the squadron in Cardiff on commissioning, and found this new broom not at all to their taste.

'They had joined the service with the intention of fighting the Germans, should war come, and had certainly not bargained for, as they saw it, being buggered about by some Army type.'

This came to a head at one early morning parade. The tension dissolved into farce when, owing to some clock changing confusion, they paraded in full darkness.

He recalled that 'good humour was restored' and that both sides recognised the need to modify their stance. The parades were eventually dropped and the senior Auxiliary pilots from Pengam fell into line.[75]

The CO's leadership style was shown in sharp contrast to at least one high-ranking visitor – Lord Trenchard, the 'Father' of the RAF, then Marshal of the Royal Air Force. JAS Watson remembered how the genial manner of 'Boom' collided with the spit, polish and parade-square lines expected for such a visit of the toppest of top brass.

'We were drawn up in a hollow square, boots gleaming, buttons shining, awaiting Trenchard's arrival and inspection at the east end of the airfield. He stepped from his aircraft and, to the consternation of those in charge of the parade, he waved everyone across to hear his message.'

He remembered, 'Trenchard was very popular and did not stand on ceremony.' This contrasted to a later visit, this time from an Air-Vice Marshal, which was altogether more frosty. 'He arrived one bitterly cold, east coast day, and insisted on inspecting the ranks as they stood to attention outside on the hangar apron near the airfield main gate, "numb" as he recalled "and frozen".'[76]

As the new CO was trying to sharpen the squadron, the nature of 614's mission began to change. Charles Bonner recalled the change in role that came with the new aircraft. 'It was decided in the light of the latest operational experience, that the squadron's role should be changed to close-support bomber squadron.'[77]

The squadron diary reported how, on 23 September 1941, they were earnestly practising bombing practice all day in the

Blenheim. The squadron took part in 'Exercise Bumper' at Odiham, its first practicing close-support bombing. In late October, further training in this new art was carried out and, in February 1942, three US Marine Corps officers visited the station to watch a close-support exercise take place for their benefit.[78]

But the squadron soon suffered a grievous blow. Flying in Blenheim IV Z6032, on 2 April, P/O John Findlay and crew (P/O George Giras and Sgt William Huggett) left Macmerry for Kinloss for a night photography sortie. After taking off from Kinloss at 22.40, no more was heard of them. Two days later, the bodies of P/Os Findlay and Giras were found on the beach near Kinloss, while Sgt Huggett was missing, presumed killed.[79]

That autumn, although the threat of invasion had receded, gas spray training continued and Britain's capacity to wage chemical war had increased. Reserves of 13,000 tons of poison gas had been gathered[80] and Britain could now take offensive action with mustard gas at five hours' notice.[81]

In May 1942, Blenheims from 614 carried out a smokescreen demonstration for Eastern Command. It was to be another display of the skill they had developed over the previous two years in gas spraying; but soon the squadron's capability would eventually be put into action – but to not to spray poison gas. The SCI's other role – for laying a smokescreen – would be needed later when the next chapter of 614's war unfolded.

The following month would see further upheaval for the squadron, but it would come not from tragic accidents or preparation and training – but from real operations. The squadron would only have a few more months before their time in Scotland was at an end, but before they left and after over two years of what some called their own 'Phoney War' – 614 were finally going to face the enemy.

CHAPTER 3

Raids (1942)

WHEN 614 FIRST saw action in May 1942, the operation they joined would herald the start of the RAF's new bombing campaign that would last until the end of the war in Europe.

The origins of this action came in the summer of 1941, with an investigation into RAF bombing – the Butt Report concluding that many raids were inaccurate and ineffective.[1] After digesting the report, the Chief of the Air Staff, Air Chief Marshal (ACM) Charles Portal, decided that Bomber Command needed a new direction and replaced Air Marshal (AM) Richard Peirse with AM Arthur Harris in February 1942.[2]

All these events would lead to a change in bombing policy, now led by AM Harris. Churchill also was clear in his desire to take the bombing offensive to Germany, speaking of his wish to 'Bomb Germany by day as well as by night... making the German people taste and gulp each month a sharper dose of the miseries they have showered upon mankind.'[3]

AM Harris had a scheme to make Churchill's wishes real – a plan to send 1,000 bombers to bomb a German city.

By 19 May he had the go-ahead from Portal who said the PM 'warmly approved'. There was not long to prepare. Harris had to assemble those 1,000 bombers from Bomber Command and Training, Coastal Command, and Army Co-operation Commands, which is where 614 Squadron sat. The aim was to have the force ready for the next full moon period in one week's time.

No. 614's task now became clearer. On 20 May, Harris wrote to the heads of the Commands explaining what he needed from them. An innovation was that bombers would fly in a 'stream' to overwhelm radar, night-fighters and anti-aircraft defences, with all 1,000 bombers to be over the target within just 90 minutes.

Harris wanted attacks to harass night-fighters and attack their bases.[4] This would be the 'intruder' force and 614 would be a vital part of it.

The need for intruders had become clear as the 'Thousand Plan' took shape. The German night-fighter force was formidable and included over 100 Me110s, alongside Ju88s and Me109s.

Harris and the planners wanted the raid to happen with a full moon and little cloud, perfect conditions for visual bombing but also for night-fighters to run amok. The night-fighters had to be kept pinned down by attack on their bases or by diversions.

The intruder force was to be made up of 50 Blenheims from Nos 18, 114 from 2 Group and 13 and 614 Squadrons of Army Co-operation Command. There would be 26 Blenheims from Nos 614 and 114 Squadrons operating from West Raynham, while the other two units would fly their 24 Blenheims from Wattisham.[5]

Their task would be to attack night-fighter airfields at Bonn, near Cologne, and Vechta in Germany; Juvincourt in northern France; St Trond in Belgium and in the Netherlands, Venlo and Twente.

The final order for the 'Thousand Plan' was issued on 26 May and said that the raid would happen on the night of 27 May or any night thereafter up to the night of 31 May, when the moon would be waning. This gave just five days when the weather would hopefully be suitable. The target was to be Hamburg or Cologne if conditions were unfavourable.[6]

It was expected to take 48 hours to assemble the bombers on the east coast. Before the order for the raid went out, Bomber Command issued instructions to move on 23 May. Movements were complicated by the need for radio silence.[7]

On 26 May, the squadron diary recorded that 'Seven aircraft left between 11.00 and 12.30 for West Raynham on detachment.' There was no indication given of why. The next day, work began to get the Blenheims ready. They were painted black, instruments were checked and bomb sights lined up. Compasses were swung and the Blenheims were flown fully loaded, guns were harmonised and fired, and in late afternoon the machines were air-tested.[8] No. 614 was ready, but before they could go the weather had to be right.

The chief meteorologist at Bomber Command's High Wycombe HQ was Magnus Spence. An Orkney man of wintry demeanour, on the morning of 27 May Spence's expression seemed even frostier than usual. Thundery conditions and heavy cloud sat over most of Germany, and Harris was forced to postpone the operation for 24 hours. The same conditions held on Thursday 28 May, and again on Friday 29 May. The window with a favourable moon was closing rapidly.

If the weather did not change, the raid could not be launched. Because of the huge numbers of aircraft having to form a constant stream to the target, movements were tightly choreographed with just minutes between each departure. The risk of collision, big enough already, would multiply. There had to be clear weather over the target for the crews to find it, and over their bases when they returned.

As the delays went on, so the pressure on Harris grew. The security risk of having thousands of extra personnel, many more movements and arrivals would grow. Normal operations and training all had to resume soon.

'There was,' Spence told Harris, 'a 50–50 chance that the cloud over Cologne would be clear by midnight.' There would also be clear weather over the bases in Britain. This was the last chance of going ahead under a full moon – if it didn't go then, they would have to wait another month.[9]

Poring over the charts, Harris traced his forefinger across a map of Europe and it came to rest.[10] He said, 'Thousand Plan tonight. Target Cologne.'[11]

On 30 May, the squadron diary noted, 'No flying took place in the morning, the time being spent on checking all the instruments, wireless, guns, etc. In the afternoon, machines bombed-up.' The waiting was almost over.[12]

That night 1,047 bombers took off to bomb Cologne, and although they wouldn't be bombing the city, 614 Squadron's Blenheims would be doing their bit elsewhere.[13]

For 614, half the aircraft had two 250lb General Purpose and 16 40lb bombs, the remainder had two 250lb delayed-action bombs and Small Bomb Containers carrying eight 40lb bombs and 24 9lb anti-personnel bombs. Their splinter effect would, it was calculated, affect morale and disturb operations. So equipped, 614 was ready to go.[14]

First to take off at 21.30 was Z5882, piloted by Battle of Britain veteran Sqn Ldr Macnamara. His aircraft joined 114 Squadron, attacking Vechta, while the remainder of 614's Blenheims would attack Twente. There was an hour's wait before the rest took off. Next was Wg Cdr Skelton and crew in M3530 at 22.30. Then, at two-minute intervals, went Peter Le Cheminant in V5451. Flt Lt Scott in V5626, F/O Robarts in R3758, P/O Maclaren in V5752. Next was P/O Smyth in BV5808 and last was P/O Baelz and crew in Z6173.[15]

The Blenheims of 614 would be penetrating 90 miles into northern Europe. Crews were told not to pass a point off the enemy coast before a given time to ensure they had full cover of darkness, and their approach to this point was as low as they dared in the failing light. Once they crossed the coast near Ijmuiden, they climbed to bombing heights – between 2,000 and 3,000 feet – and headed towards their targets.[16]

After so long desperate to see action, Le Cheminant recalled the feeling of finally facing the enemy. 'There was a beautiful full moon and I remember in particular the excitement of crossing the enemy coast for the first time.'[17]

Wireless Operator/Air Gunner (WOP/AG) Sgt Paddy Stapleton flew in Blenheim V5752, piloted by P/O Ian Maclaren and observer, Sgt Denis Ager.

The Blenheims went in low to remain undetected by German radar. The entire Dutch coast was covered by a network of 'Freya' radar stations.[18] Paddy recalled how aware they were of this. 'When we were going across the North Sea we had to go very low because of the German radar. They used to tell us that their radar could only go down to about 30 degrees, but it went down much lower than that.'

Speeding at wave-top height, they could see the spray coming off the sea. The Dutch coast came into sight. 'Then we saw the silver sand of Holland.' It was a sobering moment, 'You've heard them say, "Enemy coast ahead" in films – well, that's what they mean.'[19]

Sqn Ldr Macnamara and his crew in Z5882, having crossed the coast an hour earlier, headed for Vechta, south-west of Bremen. The weather was atrocious and visibility very bad. The 614 diary told how he and 114 Squadron failed to find the target and dropped their bombs near Nordeney Island.[20] One of the 114 Squadron Blenheims was shot down and all of its crew were killed.

Meanwhile, the rest of 614's Blenheims were en route for Twente, five miles from the Dutch/German border.[21] First over the target was M3530, flown by the CO, Wg Cdr Skelton. They failed to find it and returned to base with their bombs, landing back at West Raynham at 04.30. Later 614 attacks fared better. Peter Le Cheminant and his crew in V5451 found and bombed the target successfully. They reported no fighter opposition and little flak.

Both Flt Lt Scott, in V5626, and F/O Robarts, in R3758, had success. Flt Lt Scott's crew, Observer Sgt Johnson and WOP/AG Sgt Gifkins, saw enemy night-fighters circling the aerodrome and being signalled to land from the ground. They all followed them in and dropped their bombs. Robarts' aircraft suffered damage to its tail from flak but returned safely to West Raynham.[22]

Just as dangerous as the formidable German night-fighter force was its system of anti-aircraft defences. As well as the Freya radar network monitoring the coast, there was Würzburg radar to guide anti-aircraft guns and networks of searchlights.[23] As F/O Robarts and crew had discovered, their flak could be lethal and they had developed defences to counter low-flying aircraft. Paddy Stapleton recalled, 'The Germans had got very canny. They mounted flak guns quite high up in the air on "flak towers". If we were low they fired their 88mm guns from above you; you couldn't see them at night on these big gantries very high up.'

Many 614 crews also had to run the gauntlet of searchlights, with many caught over Holland.[24]

It was, Paddy remembered, one of the many 'hairy' moments his crew experienced on the raid. 'There was heavy AA and searchlights. We got caught in those searchlights over Amsterdam.' Paddy grimly recalled, 'They were the worst thing because once the light gets on you – it's a sort of a blue-white light and it's intense.' They did have some protection. 'You had a polaroid sun shield on your goggles like sunglasses, and you could flip them down over your eyes; they're so dense that, at night, you could use them to see right through the beam.' So equipped, they could then take the dreaded searchlights on. 'We had to dive in on those and shoot them out... Ian [Maclaren] became very adept at that and used to get us out of some very tight corners.'

Maclaren and crew found their target and started circling the night-fighter base waiting for the right moment to pounce. Paddy recalled, 'We were doing that circuit over Twente and we had not to commit ourselves because our main object was to prevent the Germans from refuelling and taking off again.'

As they circled the airfield in the dark, waiting to make their attack, they suddenly found they were not alone.

Quickly they were enveloped in a stream of Junkers 88s approaching the aerodrome. There now followed an elaborate game of hide and seek, with them hiding among the silhouettes of twin-engined night-fighters – not unlike the Blenheim. The pilot Ian Maclaren, who Paddy remembers as being 'very bold', put his own landing lights on to confuse them. Then both Paddy and Don spotted that the incoming stream of German night-fighters had put their lights underneath the aircraft out, 'so we flicked ours off'.

'I could see one coming in to land next to us,' Paddy remembered. He wanted to open fire and said to Ian, 'I can see one. Shall I have a go at him?' 'No, Paddy,' the ice-cool pilot replied. 'Sit tight and blow them a f**king kiss. If they see who we are – we're dead.' They held their nerve and they waited.

As other Blenheims had seen that night, the airfield controllers made the mistake of lighting up the airfield runway for their night-fighters. 'They lit up the flare path for him to land and we dived on the airfield behind him. "Bom, bom, bom" we went – machine-guns blazing away at hangars as we went through.' When they landed they saw that a lump had been taken out of their port wing by flak. They had been lucky more than once that night.[25]

Sixteen of the 26 Blenheims from West Raynham reported successful attacks, but the others either didn't find their targets or were forced to turn back.[26]

The results of this first 'Thousand-bomber Raid' were promising. RAF losses – at 41 aircraft – were deemed acceptable. Almost 13,000 buildings were destroyed, but the number of people killed on the ground was remarkably low, less than 500. However, over 5,000 people were injured and over 45,000 were 'bombed out' of their homes.

The raid proved that the new tactics could work. The morale of Bomber Command was boosted by this great demonstration of air power and by the wide publicity which followed. That same publicity confirmed Bomber Command's future as a major force, and it placed Sir Arthur Harris firmly in the public eye where he would remain for the rest of his life.[27]

Successful though the operation was, Harris' plan was for another raid. He was already thinking in terms of striking a double blow while the force was assembled. The disruption to training and conversion, as well as to routine operations, would last about a week. If he sent the force out two nights running, he would add 'only one day to the investment but probably double the dividend'. The idea of a double blow, either against the same target or possibly a new one, was incorporated as part of the original plan.[28] No. 614 would play their part in the next raid too.

On the night of 1–2 June, 957 aircraft were launched on Essen. This time Army Co-operation Command supplied 15 Blenheims to attack German aerodromes as the bombing force attacked the city on the Ruhr.[29]

Eight Blenheims from 614 took part with all the same crews, apart from P/O Robarts' crew which had departed to be replaced by that of P/O Porter. Twente again was the target for most of 614's Blenheims but, once more, Sqn Ldr Macnamara joined 114 Squadron for a separate attack – this time on Bonn Aerodrome near Cologne. He and his crew failed to find the target due to heavy smoke from fires in Cologne still burning from the previous night's massive raid. Instead, a 'runway of an unidentified aerodrome' was believed to have been bombed. On the way back they were attacked by a night-fighter off the Dutch coast which pursued them at wave-top height. The Ju88 kept on the Blenheim's tail, making three attacks as the WOP/AG Sgt Ryder desperately returned fire. They made their escape and landed at a diversion airfield, Swanton Morley, with no damage.

The CO, Wg Cdr Skelton, and his crew found the target and released their bombs. They too landed at Swanton Morley, but there they found that three of the bombs, in fact, had failed to drop and were still in the bomb bay. It was a problem that was to plague 614's Blenheims in the coming months.

All other crews found the target and successfully dropped their bombs. On the way to Twente, L1454, flown by P/O Porter, had to take violent evasive action to avoid heavy flak and after being caught by the network of searchlights over Amsterdam. As they flew to low level, the gunner, Sgt Petrie, reported his turret had jammed. At this point Potter decided to return to West Raynham with their bombs.[30]

The raid on Essen itself was less successful. There was total cloud cover over the city and few bombers found it, let alone hit it and 32 aircraft were missing. Harris had to disperse the 1,000-force but wanted more of these massive raids.[31]

The next was to be on 25–26 June and the target would be Bremen, home to U-Boat construction yards and a Focke-Wulf aircraft factory. Army Co-operation Command would again supply Blenheims, with twelve coming from 614 Squadron.[32]

As the 614 contingent prepared to leave Macmerry for West Raynham, one machine, V5457, piloted by P/O NM Tucker, was taxiing when its undercarriage collapsed and didn't take part in the operations. The observer, Sgt Yelland, went on to fly with the CO, Wg Cdr Sutton.

The remaining eleven Blenheims took off from West Raynham the following night between 22.40 and 00.45. Most were bound for Leeuwarden Aerodrome in the north of the Netherlands. Of the six which attacked here, V5752 flown by P/O Hannah, Z6173 flown by P/O Georges and L1454 flown by P/O Porter all bombed the target with fires seen across the airfield. Of the remainder, none definitely located and attacked the airfield but R3758, flown by P/O Smyth, dropped their bombs on a railway junction at Leeuwarden as they couldn't identify the target. V6078, flown by P/O Irving, also could not find the target, but dropped their bombs at Noordwijkerhout near the coast. Five more sought out airfields near Bremen. Three found and attacked Vechta airfield. They included F/O Ian Maclaren and crew in V5408.[33] His logbook calmly recounted, 'Aerodrome successfully bombed. Heavy AA and searchlights. Night-fighter attack repulsed.' Finally, of the two which had Ardorf as their target, one found the airfield and bombed it.

Overall, the weather was treacherous and, as with the Essen raid, the results were disappointing. Enemy production wasn't significantly hit but losses of crews and aircraft – especially those from training units – were high.

It would be the final Thousand-Bomber Raid. The massive effort could not be sustained and the results were questionable. But Harris had done what he had set out to do; he had captured the imagination of the British and American public, he had exhilarated his command, and he had won the admiration of the Prime Minister, paving the way for greater offensives to come. Churchill told him, 'This proof of the growing power of the British Bomber Force is also the herald of what Germany will receive, city by city, from now on.'[34]

The frustrations of constant training at Macmerry have already been mentioned. The interlude of the Thousand-Bomber Raids had ended in June. Would 614 see action again?

It seemed they would. In mid July, the squadron received word that they would be taking part in 'Operation Rutter'. It would see them flying from an airfield on the south coast to northern France.

On 2 July, nine Blenheims left Macmerry for Odiham.

All arrived safely apart from that piloted by P/O Smyth which overshot on landing and damaged the propellers and observer's cockpit. The crew members were unhurt.

With the news that operations were on again, Bill Hooper and the rest of the 614 groundcrew boarded an ageing, rickety transport and headed south.

'We found ourselves loaded aboard a "Sparrow", complete with tool boxes and yawing and wallowing and spewing our way to Odiham. What is a Sparrow? It's a gutted-out Handley Page Harrow, a pre-war bomber which took off at 70mph and with a top speed of little over 100mph and had precious little margin for headwinds or emergencies. You couldn't lean against the side or you'd fall through the fabric.'

One of them crash-landed at Odiham carrying 'A' Flight, he remembered, which didn't help.[35]

When they got there they found out more about what the operation would entail, but it seemed preparations for the raid were not always clearly thought out. The next day all aircraft were grounded to enable groundcrews to change camouflage from night to day. Bill Hooper grimly recalled this series of orders and counter-orders. 'First the order came to "change the colour of the underparts [of the Blenheims] to Sea Green"; then back to camouflage; then back to Sea Green. Sometimes twice a day for about a week.' Cellulose was used to mix with the paint. It was messy stuff. 'We ate cellulose, drank it, it was caked in our hair, our clothes and our skins.'[36]

The Blenheims were all air-tested on 3 July and the squadron was ordered to stand-by from 19.00. This was cancelled from 17.00 and the squadron was allowed to stand down until 19.00 the following day. On 5 July, all aircraft were again air-tested and the squadron was told to stand-down until further notice. The next day, after pilots Baelz, Irving and Maclaren took their Blenheims up for formation flying, the order came again to stand-down until 19.00 the following day.

After much painting and repainting, preparation and waiting, the news came that 'Operation Rutter' was cancelled. The detachment returned to Scotland on the same day.[37]

Bill remembered feeling despair. 'My opinion of Army and Air Force leaders was at rock bottom. I had visions of the war lasting another 20 years at this rate. Back to Macmerry we went with the sting and stench of paint remover.'[38]

The squadron would not be at Macmerry for long and would go south again when a revised plan was made. This would be 'Operation Jubilee'. This was a scaled-down version of 'Rutter' and would be an Allied amphibious attack on the port of Dieppe in northern France. Over 6,000, mainly Canadian, infantry, supported by a regiment of tanks, would be put ashore from a naval force operating under RAF protection.

The port was to be captured and held for a short period, to test the feasibility of a landing and to gather intelligence. German coastal defences and port structures were to be demolished. The raid was intended to boost Allied morale and demonstrate the commitment of the UK and Allies to opening a second front in the west in support of the Soviet Union.

By 14 August all aircraft had arrived at Thruxton in Hampshire and the squadron was settling in and waiting for orders. The next three days were filled with formation flying, radio tests and operational air tests.[39]

The squadron's job, using gas-spraying skills built over the last two years, would be to lay a smokescreen to cover the invasion forces. Preparation was not all straightforward and, as Bill Hooper remembered, there was more indecision – not over camouflage but how the smokescreen would be laid. 'The subject for indecision and confusion was bomb load. In the morning they said "Smoke bombs", in the afternoon they said "Smoke containers". Remove bombs, take off bomb doors, dozens of 2 B.A. screws and Simonds nuts. Fit the long and much more efficient smoke containers. Two hours later that afternoon, it was back to smoke bombs. Reverse the whole operation and so on through an increasingly tiring and frustrating three or four days and nights.'

After the confusion and frustration of loading and unloading smoke bombs and canisters, the squadron was poised ready, yet not knowing if the operation would ever happen. Would it be cancelled, as 'Operation Rutter' was? Would they be faced with another long journey back to Macmerry in the lumbering Sparrow followed by more training, more exercises? Hooper recalled, 'We were standing-to every morning waiting for the word to go.'[40] On 18 August the diary told how the day was taken up with 'formation practice for DIEPPE raid and time schedule trial successfully carried out'.[41]

As with the Thousand-Bomber Raid, the timing depended on the weather, but this time the tides were also a consideration. Finally, the right conditions were forecast for 19 August. On the night of 18–19 August, 237 ships sailed from Portsmouth, Southampton, Newhaven and Shoreham on the south coast and steamed out into the Channel towards Dieppe.

Commanding the air forces was AM Leigh-Mallory, AOC 11 Group, and he had put Sqn Ldr Peter Reynolds in charge of all smokescreening operations for Dieppe. He had drawn up precise plans and briefed the crews whose job it would be to lay the smoke.[42] That night, crews were told of their mission for the next day. Timing had to be perfect. Height, speed and precise landfall had to be closely watched. The diary entry for that night simply read, 'Briefing for DIEPPE raid in evening. All crews ready for operation.'[43]

The aircrews were briefed, but little was told to the groundcrews until after the operations had started.

There was real apprehension as to how the enemy would respond. As Wg Cdr Michael Pedley, OC 131 Squadron, flying Spitfires for the operation recalled, 'We were very conscious of the fact that the enemy would surely react with every fighter within call… we were very tensed up at the reaction in store for us, especially the numerical balance of forces in the air at that particular moment.'

It would be like poking a stick inside a wasp's nest. It was expected that up to 100 RAF aircraft would be lost.[44]

On the morning of 19 August, it was still dark when the aircrews at Thruxton were awoken. It was the smoke-layers – Blenheims, like those from 614, and Bostons – who would be among the first to go. Their smokescreens had to shield the landing craft from enemy fire as they approached the Dieppe beaches.

The airfield became a hive of activity, Bill Hooper recalled. 'Suddenly, in the pre-dawn gloom, we saw airmen scurrying around the ops room and on the other side of the aerodrome trucks tearing around the perimeter.' The raid was on.

Blenheims of 614's Army Co-operation colleagues, 13 Squadron would be first to go and took off at 04.15. A minute later, Bostons from Nos 418 and 605 Squadrons took off on missions to silence coastal defence guns as a prelude to the start of the landings at 05.00.[45]

At 04.17, ten Bostons from 226 Squadron left Thruxton also to lay smoke. They were joined by two Boston's, each from 88 and 107 Squadrons from Ford.[46] Just eight minutes later, at 04.25, 614 were set to enter the fray. Its first wave was divided into two formations. The first was led by the CO, Wg Cdr Sutton, in V5534. He took off at 04.25 and a minute later he was followed by the second aircraft in the formation, V6002 flown by Flt Lt Robarts. Things then began to go wrong. The third Blenheim IV, R3758, flown by P/O Hanbury, did not make it into the air. Although the diary told how Hanbury's aircraft, 'failed to take off owing to collision with car parked on runway'. One of the 614 pilots in the second wave, Peter Le Cheminant, remembered it differently. 'On the first sortie pilot officer Hanbury *did not* hit a car on the runway. He taxied into a ditch.' Either way, the formation was missing one aircraft and, after this inauspicious start, the remaining two in the formation became separated in the dark and Wg Cdr Sutton returned to base. Flt Lt Robarts and his crew in V6002 continued alone and laid their smokescreen and returned safely.[47]

Five minutes after the start of 614's abortive launch, another bid was made at 04.30 when six crews of 88 Squadron took off from Ford. It was cloudy and only three crews claimed to bomb the guns. Some aircraft were caught in searchlights but made their escape.[48]

The CO of 88 Squadron, Wg Cdr James Pelly-Fry, had just bombed those gun positions as the landing was about to hit the beach at 05.00. He recalled, 'The tracer bullets were coming from the Dieppe coastline, literally bouncing on the sea; and all in that very first light when I was returning back to Ford and flying low past the landing craft going in for the assault.' He recalled a one-way traffic system was laid on for the bombers. 'We flew in with the boats on our left and returned on the other side of the Armada.'

He remembered there was enough light to see the 'fiat-bottomed landing craft making their way'. The up-turned faces of the troops under their tin hats reminded him of a *Daily Express* Giles cartoon – 'men with snub noses and round faces'. He felt for them, thinking, 'Poor chaps, they had no idea just how tough their reception was going to be,' adding, 'I was thankful to be an air-man.'[49]

By then, the first formations of 13 Squadron had attempted to drop their smoke bombs. They had been given the task of blinding two well-positioned anti-aircraft batteries commanding the white cliffs on the left side of Dieppe. They successfully dropped their smoke bombs over the target and headed back, but one Blenheim was presumed hit by anti-aircraft fire and failed to return. The crew of P/O Woodland and Sgts Neville and Boyd were posted as missing. The RAF had taken its first casualties of 'Operation Jubilee'.[50]

At Thruxton, the Blenheims of 614's 'A' Flight waited. Groundcrew chief for the Flight, Sgt Bill Hooper, stood ready to see the aircrew off. The formation would be led by Flt Lt John Scott in V5626, with observer Sgt William Johnson and WOP/AG Sgt George Richard Gifkins.

A sandy-haired Scottish borderer, 24-year-old Flt Lt Scott had never intended to be a pilot. Before the war he studied at University College, Nottingham, where he obtained a diploma in Textiles. His father ran the Hawick-based clothing firm, Lyle & Scott (which still exists) and John's diploma was gained in readiness for a career in the family firm. However, he clearly had a thirst to serve in the military and joined the Territorial Army and was granted a commission in the King's Own Scottish Borderers in 1938. He transferred to the RAF in 1940, joining 614 Squadron in 1941.[51]

Scott was the senior officer in the formation which also included V5808, with F/O Ian Maclaren, Observer Sgt Denis Ager and WOP/AG Sgt Paddy Stapleton as crew, and T2288, with P/O Hannah, Observer P/O Campbell and WOP/AG Sgt WJ Dobson.

They were scheduled to take off at 04.30. The timetable of smoke-laying raids was set precisely to coincide with that of the troops landing at Dieppe. If there was any delay, warships, landing craft, tanks and vehicles on the beaches would be exposed to German fire.

Flt Lt Scott pulled up at his waiting Blenheim, it was, says Hooper, 'nicely loaded with smoke tanks as per the last instruction'. A look of horror came over Scott's face. 'Sergeant, what are we doing with smoke containers on?' he asked. 'We should have smoke bombs!' Hooper replied, 'Why the hell didn't somebody say so? Do they think we're bloody psychic or something, Sir?'

The tanks had to be removed, and quickly. The invading forces were due to land at 05.00 and they needed continuous smoke cover if they were to avoid being annihilated.

Bill remembered, 'Organised pandemonium broke out. Everyone piled in. No trade demarcation disputes in 614.' The SCIs were duly replaced by smoke bombs. The job was quickly done amid a frenzy of bolts being unscrewed and re-attached and bomb doors refitted. Bill later admitted, 'Those 2 B.A. screws and nuts haunt me to this day.'

The crews climbed onboard and the Mercury engines roared into life. All three Blenheims left at 04.30. Bill proudly recalled, 'The practise of the last few days stood us in good stead and the squadron took off dead on the appointed time.' The groundcrews relaxed and started to sweat it out. They had done all they could. Now it would be up to the nine men on board.[52]

With the sun rising at around 06.00, they would have taken off in the dark and would be over the target as it was just getting light. Crossing the South Downs they would fly over the English coast near Worthing, then drop down to wave-top height.

The Germans were well prepared. Opposing Leigh-Mallory's force that day, the Luftwaffe had two Jagdgruppe (fighter groups) in France, JG2 Richthofen and JG26 Schlageter. These two units had 190 Focke-Wulf FW190s and 16 Messerschmitt Me109 fighters. Bomber units, which could attack landed troops or ships, had 107 aircraft in a serviceable state. Of these, 59 were twin-engined Dornier Do217s, the rest mostly Junkers Ju88s or Heinkel He111s.[53]

Many of the first-wave squadrons carried out their missions without great impact. Neither the large gun emplacements nor the machine-gun positions in the cliffs were neutralized by air attack. This meant many of the flak batteries and machine-gun posts were ready to savage the second wave – and now they would be very awake and very alert.

At Thruxton, 'an eternity of time passed,' remembered Bill Hooper, as the groundcrews anxiously waited.

Sgt Paddy Stapleton – P/O Maclaren's WOP/AG –

remembered that as they'd learned to do in the Thousand-Bomber Raids, they flew as low as they dared to avoid German radar.[54]

The three Blenheims dropped down to wave-top height as they flew across the Channel. The beaches for the landing force were flanked by tall cliffs. This was where the gun batteries were sited which were targeted that morning by the previous raids. Anti-aircraft guns also placed on those cliff tops were all very much still firing.

Then, as the formation approached the target area, a burst of fire hit the lead Blenheim. Flt Lt Scott, in the pilot's seat, bore much of the force as fragments of shrapnel tore through the cockpit. The gunner F/S Gifkins in the turret was cruelly exposed and, too, felt the full explosive force of the shrapnel. He was killed instantly.

The windscreen was shattered and the instrument panel and compass destroyed. The white-hot metal tore into Scott's left leg, both arms and his face.

He tried to tell his formation to continue to the target but found he couldn't speak, so badly injured was his face. His jaw was almost severed and held on only by the straps of his helmet. In any case, the aircraft's radio was knocked out, so he was unable to communicate with the rest of the formation. In unimaginable pain, his observer, Sgt Johnson, managed to help him keep control of the aircraft.

The damage to his machine also affected the bomb release gear, so despite his efforts he was unable to release his smoke bombs. He turned and headed back. The other two aircraft following the formation leader and maintaining radio silence, as they were ordered to, did not drop their smoke bombs and also headed for home.[55] Bill Hooper recalled that in accordance with orders on maintaining strict wireless silence, all sets were switched off and only the formation leader was in wireless communication with base. The orders for the remainder of the formation were to conform to the leader's movements or voice control on VHF radio.[56]

From the squadron, only one aircraft of the first two formations had been able to lay any smoke on the enemy shore. One of its aircrew was dead and one was badly injured. Could the next wave carry out their mission? They were set to take off between 07.35 and 07.40.

Just half an hour earlier, six Bostons left Ford to bomb the 'Rommel' battery on the cliffs. Over the target about 20 Fw190s swooped in from all directions to attack the unescorted bombers. One Boston had to ditch after being hit by a shell and an explosion. The two gunners bailed out and survived, but the pilot and observer were lost.[57] It was clear then what kind of reception awaited the following 614 Blenheims over Dieppe.

The next formation from 614 would be made up of Blenheim V6002, with Flt Lt Le Cheminant and his crew; V5534 with P/O Georges and his crew and T2288 with P/O Baelz and his crew.[58]

Le Cheminant came to rejoin 614 and take part in the Dieppe raid via a series of coincidences. He had been posted to No. 2 School of Army Co-operation at Andover on 14 June but was posted back to 614 Squadron on 10 August. He rejoined them at Thruxton where, he recalled, 'preparations were being made for some major operation which turned out to be the Dieppe raid.' He had no crew but one of the pilots went sick the day before the operation and he replaced him.

A native of Guernsey, Le Cheminant was astonished to find one of the crew he 'inherited' was a fellow islander, Observer P/O Leonard Quevatre. This was an incredible coincidence considering Guernsey had a population of just 40,000 in 1940. In a further Gallic twist, the WOP/AG, Sergeant Limoges, was a French-Canadian.

By 07.40 the sun was up, so there would be no cover from darkness for the three Blenheims. They lifted off and headed for the Channel. Seven minutes before, Bostons from 88 Squadron set off again at 07.33, sending six more aircraft to attack another battery, but again their bombs missed. A desperate plea for more smoke came from the land forces at 07.00.[59] Would 614 be able to answer?

As they approached the French coast, smoke would already be visible. The sea would have been thick with warships and landing craft. But before they could play their part, once again 614 would be frustrated. For reasons unknown, the sortie was called back. At this point again, Le Cheminant is again at odds with the squadron diary. While the official record states that the formation was recalled by 'R/T' (radio transmitter). Le Cheminant states that his log book recorded that they were turned back, not by radio, but rather by Spitfires from their escort diving across their bows. Disconsolate, they landed back at Thruxton by 09.00.[60]

As Le Cheminant's formation touched down, preparations for the next smoke-laying missions were underway at Thruxton and at Ford. In their second smoke operation, four crews of 226 Squadron left at 10.15 to blot out the aim of the 'Hindenburg' battery. They attacked at 11.13, leaving a good smokescreen which drifted seawards. At 10.15 two other crews of 226 Squadron, fitted with smoke canisters, left to place a screen over the Dieppe beach, trailing it from 50 feet for a mile or so. This was the most hazardous operation so far, as they found when speeding through crossfire from the enemy above on the cliffs and 'from untrusting friends [the Royal Navy] at sea'. Of the smoke-layers, two crash-landed, one at Shoreham with a wounded gunner and the other at Gatwick, also carrying casualties.

At 10.50, 614 launched its third and biggest sortie with two formations – one of three aircraft and one of five.

The first was led by P/O Smyth in V6078, with Z6104 flown by P/O Georges, and Z6173 flown by P/O Porter. They were supposed to get fighter support from a rendezvous point mid-Channel. Although this escort failed to turn up, all three Blenheims made it to Dieppe and successfully laid their smokescreens. The whole formation returned safely between 12.50 and 13.05.

The second, larger formation took off at 10.50. It was led by the CO, Wg Cdr Sutton, in V5534, and was to have had six aircraft. He was joined by Peter Le Cheminant and his crew in V6002, P/O Hanbury and his crew in R3758 after his earlier crash that morning, P/O de Wesselow and his crew in R3536 and P/O Baelz and his crew in V5808. P/O Hannah and his crew, presumably in T2288 which they had already flown in, also took off at 10.55 but returned five minutes later due to engine trouble.

A most graphic and moving account of that sortie came from de Wesselow's WOP/AG, New Zealander Sgt Fred Bunting, who was joined by Observer Sgt Ivor Self. Fred recalled that the original plan for 614 was to lay a smokescreen over the beaches by about 17.00 to allow the troops to withdraw, but by mid-morning it seemed that the raid was not going well and they would be needed long before then. 'We were now required to do the job much earlier at around 11.00. Our eight Blenheims, flying in two groups, were allocated an escort of fighters. However, because of the problems in starting the engines of one aircraft in our group [Hannah], we were delayed in take-off, missing the fighter cover.' Their formation would be on their own.

Soon after leaving the south coast, they were attacked head-on by two Me109s, but their luck was in. 'After the initial pass, they decided to quit – perhaps short on fuel or ammunition, or the end of their patrol… it was very unlikely that our outclassed armament, or perhaps the profanity directed at them, drove them off.'

Unlike earlier 614 Blenheims, they carried smoke canisters. To lay the smokescreen they had to fly alongside the cliffs, providing a perfect target for guns hidden in the cliff sides.

'We swung into the town at rooftop height, laying the long curtain of smoke to drift down onto the beach – hopefully giving cover to the troops and the landing craft.'

Fred then saw why they had been called to lay smoke much earlier than planned. He paints a picture of a hellish, surreal and violent few minutes. 'The racket of gunfire was intense, even over the noise of our own aircraft.' As they laid their smokescreen, they flew through virtual curtains of flak – exploding shells filling the sky with deadly shrapnel. He recalled 'blue, black and red streaks, complete with black exploding balls of shrapnel – a bit like flying through a giant spider's web'. They were hit many times, as were the other aircraft in the formation, but somehow they flew on.

As a gunner, it was Fred's job to watch for attacking aircraft. As he anxiously scanned the skies around them, he was presented with a bizarre scene unfolding below them as the Blenheim thundered along at roof-top height. 'I remember almost unreal snapshots of smoke everywhere – chimney pots and a church spire flicking past.' He had a grim view of how the landings were faring. 'I saw dozens of bundles on the beaches (at the time I couldn't believe they were bodies) – knocked-out tanks – beached or burnt-out landing craft, and the skies full of aircraft.' He recalled, 'Nothing seemed to be moving on the ground, but some small boats were taking a hammering.'

Having laid their smokescreen they left this harrowing scene behind and headed out to sea. 'It was back to wave-top height again and heading home as fast as possible.' They sped away – desperately trying to outrun the furious barrage still being launched at them. 'There were heavier calibre shells still splashing in the water around us.' They hit one of the Blenheims, but still their luck held and the wound was not fatal, although as Fred remembers, 'it left one of our group with a large hole in its rudder'.

All shipping was given a wide berth on the way back to

base. 'The Royal Navy, frequently poor at aircraft recognition, would obviously be more trigger-happy on this day,' he recalled.

They glimpsed the white cliffs of the southern English coast and he admitted, 'they never looked better as we approached them'. Landing back at Thruxton the crews waited outside the operations room to see if they would have to go back. It had, he remembered, 'been an extremely unhealthy trip'.

It was his first real operation, and although he had come through it alive, he admitted he was beginning to believe the rumoured statistics about aircrew and wondered if he would reach his 20th birthday.[61]

In the same formation, Peter Le Cheminant too recalled the hail of fire that met them. 'We came over the beach low in close line astern and received intense fire from the cliff top. I was directly behind [Wg Cdr and CO] Harry Sutton and could see his aircraft being hit repeatedly in and just forward of the tail section. I, in turn, took most of my five major hits in that same area, as did the following aircraft. I think we were saved from taking our major damage in the cockpit area by the fact that we had been preceded over the target area by Bostons of 226 Squadron dropping at 300mph, whilst we dropped at 180mph. This would almost certainly have caused the Germans to allow too much deflection, with the result that we were probably being hit by fire intended for the aircraft immediately astern of us. Either way, it was a brisk and energising few moments. We split up and came home individually at low level.'[62]

Bill Hooper and the groundcrew, having seen the formation off, were waiting. 'Slowly they came straggling back,' he remembered, 'not as a squadron, but ones and twos.' It was clear what they had endured and how lucky they were to have survived. 'Thank God they all came back, but in what a state! Bullet holes everywhere. It seemed to me that most of the bullets had entered from above showing that they had been flying below the level of the Dieppe cliffs as they laid their smokescreens over the beaches and the Germans had been firing down on them.'[63]

They had been lucky indeed, but there might have been other reasons for their escape that day.

It was later revealed that Wg Cdr Sutton's gunner, P/O Sloan, played a key role in preventing the four Blenheims following them being hit by the torrent of anti-aircraft fire. The Blenheims of 614 had to fly low – at 50 feet – and keep steady as they laid their smokescreen. This made them very vulnerable and it was difficult for the pilot to take evasive action when they were being fired at. But, from the CO's Blenheim, P/O Sloan kept firing into the flak positions, suppressing their fire just enough to prevent them scoring fatal hits on the aircraft following them.[64]

Earlier, the formation led by John Scott's stricken Blenheim had limped back across the Channel.

Badly wounded and feeling himself rapidly losing consciousness, Scott nursed his damaged Blenheim back, leading the two other aircraft in the formation to the English coast.

RAF Friston in Sussex was home to the Hurricane fighters which operated from there during the Dieppe raid, and was also used as an emergency landing ground for the operation. From far out at sea, the nearby towering cliffs of Beachy Head would have acted as a beacon for the badly-injured Scott.

As the coast came nearer he would have seen just to the left of Beachy Head and inland the airfield at Friston, less than a mile from the Sussex coast on the South Downs. This was where Scott chose to put his wounded Blenheim down.

Also in the formation, Paddy Stapleton recalled how, after John Scott peeled away to make his emergency landing, Ian Maclaren led the group – now just his and P/O Hannah's Blenheims, back to Thruxton.[65]

By this time, Scott would have been losing blood and

struggling to stay conscious. His smoke bombs, still on board, were likely to have contained phosphorous which, as well as generating smoke, created incredible heat and would burn terribly.

Unable to jettison his bombs, the landing would have to be as smooth as he could make it if they were not to detonate. The airfield at Friston had a grass runway which would help soften the impact of a landing, especially if there was a risk it might be hard one.

But, as he descended, his problems, already considerable, mounted further. His starboard engine began to lose power. Preparing to land, he found that his hydraulics too had been destroyed and so was unable to lower the flaps. He would land much faster than normal. Without hydraulics, the undercarriage would not lower. He would have to make a belly landing, meaning the bomb bay, carrying those smoke bombs, would bear the full force of the impact.

Scott brought the Blenheim down as gently as he could manage. Adrenalin would have been coursing through his body since the shell fragments tore through him over the French coast. Did he have any strength left? He was about to find out.

The Blenheim approached the ground, the grass airstrip quickly looming through the smashed windscreen and glazed nose. Sgt Gifkins' lifeless body would be behind him in the turret. How much warning he and Sgt Johnson got of the impending disaster is not clear. There would have been mere seconds between the impact of landing and the roar of explosion. The contact with the ground detonated the smoke bombs and the Blenheim was engulfed in flames and acrid, burning smoke.

Sgt Johnson was horribly burned and now unconscious. Scott, although seriously wounded, and now badly burned himself, somehow struggled free and pulled the unconscious Johnson clear, before rolling down the Blenheim's burning wing to escape. Both he and Sgt Johnson were taken to the Queen Victoria Hospital at East Grinstead where they were treated for terrible burns and shrapnel wounds.[66]

The groundcrews had continued their vigil at Thruxton, anxiously watching the skies.

They waited and they waited, until they caught sight of the aircraft coming to the circuit to land. Bill Hooper remembered watching as the Blenheims returned, 'A swift count. One missing. Who's got the chop? Oh Hell, it's Flt Lt Scott, our "A" Flight commander who had been leading the formation.

'The aircrews get out from the planes, a babble of excited voices and bewilderment on every face. Consternation. Nobody had laid smoke. The beaches on one side of Dieppe were not masked out, the German gunners had a clear view and slaughter could commence. Where was Flt Lt Scott? We saw him crash-land someone answers. The sense of dismay and disappointment is overpowering.'[67]

The CO, Wg Cdr Sutton, went to East Grinstead to see the injured crew as did Ian Maclaren. Horribly burned, Sgt Johnson died of his injuries after 36 hours in hospital.

Maclaren, flying to Friston in an Airspeed Oxford went to the hospital that day.[68] Paddy Stapleton recalled, 'When he got there the navigator was still alive, but he was badly burned.' Paddy believed that it was what would become known as a 'friendly fire' incident and that the Royal Navy were to blame. A later report told how they were hit by fire from 'an armed ship',[69] but, as Paddy hinted, they ran the gauntlet not just of enemy fire but shells from the Royal Navy.

The Dieppe raid was quickly turning into a fiasco. The plans to knock-out the heavy gun and flak batteries around the port had failed. Too many of the smokescreen-laying missions had also failed, partly because those anti-aircraft guns had shot the smoke-layers from the sky. The Churchill tanks of the landing forces had become bogged down on the steep, shingle beach and had been shot to pieces – as had the troops landing alongside them. All that could be done now was to withdraw.

Among the confusion and carnage that day, two former members of 614 Squadron had an unlikely reunion. S/Ldr Richard 'Dickie' England was one of the first intake of trainee pilots at Pengam Moors in 1938. He was temporarily stationed at RAF Ford in Sussex at the time with 88 Squadron, and led formations of Bostons which joined 614 in laying smokescreens.

Dai Park, as an 18-year-old AC2, who also had joined the squadron in 1938, had been batman to the then P/O England when they were stationed at Inverness in Scotland with 614 in 1940. He recalled the disappointment at being 'shanghai-ed' into the role, being tempered by being allocated to the likeable officer. He confessed his heart wasn't in the job, with 'spanners and oily rags' really being his 'forte', but one encounter with him stayed with Dai for many years after. 'The talking-to he gave me on the subject of personal values, I believe, instilled in me the real value of patience and ambition if you want to reach your ultimate goal.' This Dai took onboard. The next time they met he had become a motor transport fitter and was now a corporal. That meeting happened just after Sqn Ldr England had returned from a sortie over Dieppe. Dai had left 614 Squadron in early 1941 and now was stationed at Ford and recalled, 'He greeted this particular Corporal MT/Fitter as an old friend, such was the stuff this man was made of.'[70]

By 22 August all had returned to Macmerry and a 48-hour pass was granted in recognition of the 'good effort on the Dieppe raid'. But the squadron would only have days left in Scotland and on 24 August it received a signal saying it would be moving to Odiham.

The story of John Scott would follow the squadron when it was posted far overseas to the deserts of North Africa six months later. After he had recovered from his injuries, he would be awarded the Distinguished Flying Cross on 22 September 1942 for the courage he showed in continuing to lead his formation despite his injuries, and for rescuing his injured observer from the wreck of his Blenheim.

Another DFC went to P/O Dennis Sloan on 2 October 1942 for his skill in drawing enemy fire away from the other aircraft in his formation.[71]

Although distraught at the terrible injuries Flt Lt Scott had suffered and at the death of his crew, Bill Hooper recalled the story did not end on that tragic note.

'The following May, and over a thousand miles away in the vast uplands of Algeria at a place called Oulmene, the North African campaign was over and we had played our part. We were told that the "Top Brass" were on a tour of inspection and we were to look busy.

'A gaggle of High Brass approached. I turned my back and pretended to be busy. Suddenly, a strong, well-remembered voice hailed me. "Hey! Sergeant Hooper! Are you still bloody-well psychic?" I turned to face a tall, sandy-haired Squadron Leader holding out his hand for me to shake. Giving the best and most willing salute of my service career and ignoring the stares of the "Brass", I shook hands with Mr Scott ex-"A" Flight commander and leader of 614 Squadron on that eventful day. The lower part of his face was covered in healed-over scars but that gunfire over Dieppe had certainly spoiled his good looks.'[72]

CHAPTER 4

North Africa (1942–1943)

WHILE THE WAR in Europe ground on with the Thousand-Bomber Raids and forays on the ground like Dieppe, the campaign in North Africa gave more than a glimmer of hope.

Following the Battle of El Alamein in November 1942, the Allies had gained the upper hand and began to roll the Afrika Korps back away from Egypt.

To increase pressure on the Axis in North Africa, forces would be landed on the Atlantic coast in French Morocco and in Algeria with a view to push Rommel's forces from the west as well as the east. Thus, 'Operation Torch' was conceived.

With the squadron blooded by the disastrous Dieppe raid, it had moved to Odiham in August to await its next posting. Although in the early autumn they didn't yet know it, 614 Squadron would be finally leaving home shores to fly to Algeria and play their part in this great operation. And they would be getting new aircraft.

Sometimes described as the Blenheim Mark V, the Bisley was broadly similar to its Mark IV predecessor which 614 had flown since 1941. With some additions intended to improve it, that did anything but.

Air and groundcrews set about doing what they could to remedy its worst fault – excessive weight. This was caused by a power-operated turret, additional armour plating and a large, heavy system of mirrors enabling the navigator to fire two .303s rearward from the nose. While at Odiham, groundcrews were ordered to remove the mirror system and some of the armour plating. The turret, at least, was seen as an improvement on the Blenheim IV.[1]

WOP/AG Sgt Fred Bunting, however, was pleased to see some new features. 'We would have a bit more speed and more armour underneath, but probably more welcome was the updated radio equipment, and the modern 360 degree traverse gun turret that we could have done with at Dieppe!' There were echoes of Dieppe when, on 29 September, there was a party to celebrate the award of DFCs to Flt Lt Scott and P/O Sloan.

Peter Le Cheminant, however, was clear in his assessment of their new type, calling it 'an abortion, a cobbled-up disgrace to Bristols, not fit to operate in the air environment of late 1942 and inferior in almost every way' to the Blenheim IV that it was replacing.

Concerns over aircraft aside, it still wasn't clear where their destination was; but there were clues when they were issued with tropical kit, including a pith helmet.[2]

Although soon to be deployed far from their home in south Wales, the links with Glamorgan remained strong. On 8 October, Wg Cdr Sutton and a crew flew to Cardiff to visit Air Commodore Cadman (who had been made the squadron's Honorary Air Commodore when he ceased to be the CO in October 1939) and 'collect various sports equipment games for the airmen to use overseas,' as well as presenting the CO with twelve bottles of whiskey.

These weeks brought excitement, as preparations stepped up, but also tragedy. Aircrew practised formation flying and D/F (direction-finding) loop calibration, necessary for long-distance flight.

South African pilot Lt Gunnar Norenius was returning to Snailwell in Cambridgeshire from a cross-country training flight, on 11 October, when it seemed that he mistook a green signal light on a nearby railway line for that of the airfield approach and descended too soon. His Bisley crashed into some trees on the airfield perimeter. The now Sqn Ldr Charles Bonner was Flt Cdr 'A' Flight and had already touched down at Snailwell. Norenius was flying with Bonner's navigator/observer, P/O Albert Drake, that night. Both were killed. WOP/AG Sgt Russell escaped through the broken back of the aircraft and, although badly burned, Bonner was quickly at the scene and gave him first aid.[3]

By this time, unbeknown to 614, their destination was clear and the Commander of the Wing under which they'd operate had received instructions. Stamped 'MOST SECRET', the orders were given to OC 326 Wing, Group Captain Laurence Sinclair GC, on 16 October, telling him how the Wing was to prepare and operate as part of 'Operation Torch'.

The instructions read, 'Your Wing will be based initially at Blida [in Algeria]. Thereafter, the Wing, or part of it, may be called upon to move to aerodromes to the eastward in rear of the Army's advance.' The Wing would be made up of Nos 13, 18, 114 and 614 Squadrons.

The orders continued, 'All four squadrons must be prepared to operate as general purpose squadrons.' Their duties could be any of the following: bombing; close or direct support of the Army; attacks on shipping; anti-submarine patrols; photographic or visual day reconnaissance and night bombing.[4]

Soon after, Sinclair visited the squadron and told them all he could about their approaching move. Le Cheminant recalled the inspirational effect of his talk. 'He enthused us with the prospect of the adventure which lay ahead. The knowledge that we would be going to war, 72-aircraft strong, somewhere overseas, was exciting.'[5]

On 1 November, the CO and Flt Lt Wallace went to the Air Ministry 'regarding the move', where they were briefed on the instructions Sinclair had received. On the same day, airmen packed equipment into lorries to be taken to nearby Winchfield station, and two days later all had been loaded onto trains. Final preparations and farewells were made on 6 November. Aircrews went on a 48-hour pass and the groundcrews and staff departed for the Clyde where they would board a troopship.

On 16 November, the aircrew flew to Portreath in Cornwall, their staging post for the long flight south. They then prepared to fly to Gibraltar early the next morning.[6]

The 'Operation Torch' landings in North Africa began on 8 November. While US troops had sailed across the Atlantic to land on the Moroccan coast, British forces landed in Algeria. It was only then that aircrews learned their destination was Blida Airfield, just west of Algiers. On 17 November the squadron began its long and perilous flight towards action. The route chosen was due south over the Bay of Biscay, turning right over northern Spain, then crossing Portugal, then to Cape Trafalgar, and finally Gibraltar. After refuelling, they would make for Blida. None of them had flown this distance before and, heavily laden with extra fuel, they were heading off into the unknown.

Half of the squadron took off for Gibraltar in the early hours of the morning, with the remainder to follow by day. All bar one took off safely. Flt Lt Robarts taxied into a barbed wire fence at Portreath and was to follow on the next day after a new propeller was fitted.[7]

Paddy Stapleton recalled his crew took off from Portreath at 03.30 in Bisley BA809. Leaving a wintry Cornwall in the early hours of a November morning and dressed in warm clothing and sheepskin Irvin flying jackets, the heat of Gibraltar felt almost tropical as they stepped onto the tarmac. Their stay on the Rock was short, and the next day they left for Algeria.[8]

Crews flying south had to watch for prowling German

Ju88s over the Bay of Biscay. But this was not the only concern for some flying the long, lonely flight to Gibraltar.

One 614 aircraft did not make it. The squadron diary noted on 18 November, 'All those dispatched from Portreath arrived in Gibraltar in good time with the exception of P/O de Wesselow and crew who were reported as missing.'

He and his crew took off at 04.00 from Portreath on 17 November and there was no hint as to what happened to de Wesselow and his crew, F/Sgt Ivor Self and Sgt Fred Bunting.

Peter Le Cheminant feared being intercepted by Ju88s on the flight to Gibraltar. He saw none of these but was buzzed by a Portuguese fighter. Had de Wesselow's Blenheim been shot down? Or had they plunged into the sea as other Bisleys from other Wing squadrons had done on their hazardous journey south? Their fate would be revealed much later.

After the war Fred Bunting told how, having departed Portreath, the flight was going to plan. 'Our take-off (or rather "stagger-off" with extra fuel tanks) for North Africa was uneventful, but with complete engine failure at about 8,000 feet over Portugal, the Bisley force-landed in the countryside, ploughing up someone's field in the process.' Bunting paid tribute to his pilot. 'We were, without doubt, being saved by Peter de Wesselow's skill.' Described as having a rower's physique, the former guardsman's strength undoubtedly helped bring the stricken Bisley safely to a halt in a Portuguese field.[9]

They force-landed at about 10.00 near Aldeia Nova in northern Portugal.[10] Bunting recalled how they were escorted from the crashed Bisley by Portuguese soldiers and civilians, one of whom was carrying a double-barrelled shotgun. They were taken further south to the small medieval town of Elvas, where they were interned for a few months with a Portuguese cavalry regiment. They stayed in Portugal until January 1943 when they were repatriated via a Dutch KLM flight to London, which was still operating.

While the aircrews had to brave the long, uncertain flight south, Ju88s and internment, they at least would reach North Africa quickly. A longer journey faced the groundcrews. With their equipment they boarded the troopship *Strathaird* on the Clyde. Like the aircrews, they didn't know where they were going but news came as they sailed. Sgt Bill Hooper recalled, 'On the morning of the 8th, the tannoy burst forth with the news that British and American forces had landed in North Africa.' They discovered they would be part of the Allied effort there.

The sea journey was a hazardous one. The *Strathaird* formed part of a protecting convoy as it sailed across the Bay of Biscay, on whose coast were U-boat bases in occupied France. Hooper recalled the flurries of anti-submarine activity when depth charges would be fired by escorts and 'monstrous mushrooms of water' would erupt.

The liners, stripped of luxurious fittings, had basic accommodation, Bill remembered. He and other NCOs had cabins (albeit often sleeping on the floor) while other ranks had to make do with the mess decks.

The airmen were outnumbered by the army, but a battle of wills ensued between those in light-blue and the Army warrant officer in charge of Bill's deck. The airmen wore ties beneath their working dress which the Army W/O took against and, when their deck mustered for parade one morning, he ordered them to remove them the next time they formed up. As Bill remembered, the airmen decided to stick to their regulation dress. 'On the next parade, no-one took a blind bit of notice and he went berserk. He threatened us with all sorts of punishments but we stuck it out and remained dressed with collar and tie.' His temper was eased slightly when the airmen had to change to the khaki battledress the army wore.

Aside from watching the escorts hunt submarines and baiting the army, the airmen played bingo and cards. There were times when their journey south resembled the voyages the ship would have made in peacetime. Bill recalled an evening when, against the backdrop of a breathtaking sunset, he and his pals watched a school of whales surface alongside the ship.

Another evening, Hooper was joined by the squadron's medical officer, Flt Lt Davies, at the rail, who said, 'You know, Sergeant, you'd pay £5 or £10 a day for this in peacetime.' How distant those days must have seemed.

Eventually they were thrilled to see the distant glow of a city to starboard. Tangiers was brightly lit, while a darkened Rock of Gibraltar glided past on the port side. The escorts continued to battle the submarine threat, and Bill noted, 'we didn't get much sleep for the next couple of days'.[11]

In Algeria, combined US and British Commando forces tasked with taking Algiers met French resistance after landing in the early hours. There was uncertainty how Vichy French forces would react, and their fighters battled Allied aircraft during the landings but some French forces sided with the Allies.

Blida Aerodrome was a French Air Force base about 28 miles south of Algiers, with a single runway, hangars and some barrack blocks. Another squadron in 326 Wing, No. 18 Squadron had arrived on 14 November and their aircrew and Bisleys were already in place.

Although a ceasefire had been agreed between Vichy French forces, with many helping the Allies take control of Algiers, the end of French resistance did not mean the Allies were safe.[12] Blida had only recently been captured, so Paddy Stapleton and his crew, feeling vulnerable as they flew over territory so recently held by the enemy, let their wheels down to show non-aggressive intent.[13]

The first five Bisleys arrived on 17 November, with the remainder of the aircraft joining them the day after. Three aircraft were missing, however. P/O Marston's Bisley was unserviceable at Gibraltar, P/O Aubrey-Rees remained sick at

Odiham, and there was P/O de Wesselow and his crew who'd never make it to Blida.

Peter Le Cheminant and his crew, Observer P/O Bill Service and WOP/AG P/O 'Ginger' Ryder, were greeted, while clambering down the wing, by a cheerful groundcrewman saying, 'Lost five out of twelve yesterday, Sir,' referring to a disastrous attack by 18 Squadron on Bizerta Airfield. He wrote in his diary, 'Blida populated with fairly friendly Vichy – the usual ropey l'Armée de l'Air types and their still ropier aircraft. Aerodrome set in a basin surrounded on three sides by hills, with mountains rising sheer to the south. The atmosphere simply reeks of the Foreign Legion, very colourful with the Tricolor still flying.'[14]

The aircrews of 614 were quickly in action. As Axis forces were pushed towards Tunisia, Bizerta became their main source of reinforcements and so was a key target. Just a day after all the squadron's aircraft arrived, a message on 19 November told of shipping targets in its harbour. Group Capt Sinclair called on 614 Squadron for seven aircraft for attacks on them and again on Bizerta Airfield.

With their own groundcrew and stores yet to arrive, 614 relied on 18 Squadron and RAF Servicing Commandos – groundcrew trained as soldiers – to arm and maintain their aircraft. In return, they loaned three Bisleys to 18 Squadron to replace those lost on the journey south.

The Bisleys were quickly prepared. The squadron diary reported, 'Aircrews brought aeroplanes up to scratch and removed all unnecessary equipment. Bomb racks were fitted and bombed up by 18 Squadron engineer officer and groundcrews. Wg Cdr Sutton, Sqn Ldr Bonner, Flt Lt Maclaren, F/Os Arthur, Georges, Smyth and Tucker and crews would fly. All seven returned safely before breakfast having carried out their mission successfully. Aircraft refuelled and re-bombed to be in immediate readiness for further attacks.'[15] No. 614 had announced their arrival in North Africa.

Although it was some days after the French surrender that the *Strathaird* docked at Algiers, troopships were constantly under heavy air attack. The diary told how, on 21 November, groundcrews and equipment 'appeared' to have arrived at Algiers, and 'nightly get the attention of the Luftwaffe', adding, two days later, 'our troops have had a rather bad time by continuous night attacks by the enemy'.[16]

That day, another troopship, the *Scythia*, moored alongside the *Starthaird*, was hit and sank, the impact nearly capsizing the *Strathaird* too. Bill Hooper recalled the ensuing chaos. 'My mate and I slid down the floor crashing against the ship's side. All the lights went out and the ship slowly righted herself and emergency lights came on.' Wing-mates No. 13 Squadron lost all their kit and some of their airmen.[17]

The men of 614, and with hundreds of soldiers, disembarked quickly, fearful of another attack. In the haste to get the *Strathaird* away, it left with most of 614's equipment still aboard. What had been unloaded was then damaged by another air attack.[18] This was reported when the CO and two officers went to Algiers to locate unit equipment and the rest of the officers' baggage, finding 'only about ten per cent of unit equipment and no officers' camp kit'.

Having disembarked, Bill Hooper, with two corporals and ten airmen, was ordered to stay on the quay, clean all the squadron kit, and guard it until relieved. They slowly gathered the equipment and he recalled how, 'it was all a bit chaotic as stuff had got a bit mixed up when the ship had listed so badly.' Hooper and his team stacked the boxes of stores, which together with the sea wall provided 'a very smelly shelter when things got too noisy'. Huddled there on that first night on North African soil, they listened to the gunfire from ships anchored in the harbour and from the city. It was a sound they quickly became familiar with, he remembered.

The next day they borrowed a boat from a local and

retrieved boxes of stores which had been flung from the *Strathaird* into the harbour during the air attacks.

Finally arriving at Blida town by train on 27 November, they marched to the airfield, arriving at 03.00. Two hours later some were back on the train as part of a 50-man advance party to Setif, 186 miles to the east, led by Sqn Ldr Bonner and P/O Bebb. Among them was the just arrived Sgt Bill Hooper. Bonner's orders were to take over the airfield if they found it to be suitable for operating Bisleys.

He recalled the journey to Setif. While slow, uncomfortable and being vulnerable to attack, they had plentiful hot water from the steam engine pulling their train. This gave them many brews and water to wash and shave. They were grateful, then, for the ever-present empty four-gallon petrol tins. 'RAF personnel lived for months on those tins,' he remembered. 'They were used as seats, ovens, baths, latrines, runway markers, and even stepping stones whenever we met up with quagmires, or similar.' Also used to build makeshift shelters, he recalled, 'they gave the groundcrews problems later on, as Sgt Bill Hooper could tell you. He and his crew had to fish dozens of the caps used in sealing the tins out of the aircraft fuel tanks.'

Conditions in Blida too were spartan and many had to make do with tents as there was not enough space in the barrack blocks.[19] LAC Andrew Turnbull of 614 Squadron recalled the shocking state of the latrines. 'We tried to open the door to the latrine but found it wouldn't move because of the thick layer of filth dried on the floor.' The Army built some 'Elsan-type latrines with crude wooden seats', which helped.[20]

Raids then continued by day and night against enemy airfields and harbours in Tunisia. Among the crews was that of Paddy Stapleton WOP/AG, with pilot Flt Lt Ian McLaren, and Observer Denis Ager. He recalled an early raid following the initial attack on Bizerta which didn't go quite as planned.

'This particular night the clouds came right down, lightning flashing, you could see the white tops of the sea. We were a bit off-track.' They saw *a* harbour, which, Paddy recalled, 'was all lit up, the ships were all there and they're off-loading and everything. We thought, this is a gift from paradise.' So they attacked.

'I'm blazing away, straight down the middle, we dropped our bombs and we did what we thought was a glorious job. We scattered them, the lights have gone out, we'd dropped bombs in amongst the ships, there's fires galore, lovely. Oh, a VC for everybody.

'So, we come flying home, full of elation and when we got back, Sinclair says, "You bastards, you weren't supposed to do that. Do you know what you've done? You've bombed Tunis." We said, "Well, isn't that all right? That's where Rommel is." He replied, "Yes, but you were supposed to be doing Bizerta, round the corner."

'They didn't want to do Tunis yet – they were going to make a daylight [raid] of that and catch the whole lot and we'd spoilt the whole thing by going in there early.'[21]

Opposition was fierce and the squadron's first casualties came on 2 December. Among the Bisleys which took off from Blida at 02.15 to attack Tunis Aerodrome was BA872, piloted by F/O Thomas Young with Navigator W/O1 George Campbell, and WOP/AG Sgt A Walsh. They had not returned.

The Wing quickly settled into a routine of attacks on enemy targets in Tunisia, often responding to direct requests from the Army. The front line was moving east and 614's original home at Blida was already 500 miles west of their targets when they arrived there in mid-November, hence the departure of an advance party to Setif. By early December, some of the squadrons in the Wing had moved to Canrobert in preference to Setif, around 300 miles further east, with 614 set to follow. They were often dispersed to Souk El Arbaa, 125 miles still further east, as a forward base where they would await the Army's calls for support.

Friday, 4 December 1942, would be a momentous day for 326 Wing and it would see its squadrons cruelly exposed in flying aircraft that were not up to the task in daylight without fighter protection – and with catastrophic results.

The boss of 18 Squadron was Wg Cdr Hugh Malcolm, a talented Scot who was made CO aged just 25. That day he had already led a flight of six Bisleys in response to a call from the Army to attack positions in the Chouigui area, between Tunis and Bizerta.

Lt Gordon Wallace, a Royal Navy Observer, seconded to 18 Squadron, recalled how they'd attacked an enemy airfield six miles east of Mateur earlier that day. They returned at 14.00, having stumbled on, as Wallace recalled, 'a lion's den… but fortunately the lions were caught unawares'.[22]

The crews had been in the air for three and a half hours and would have been tired. Yet, on their return, Malcolm agreed to a request from the Army to attack the same airfield where the lions were not likely to be caught napping again. The other squadrons in the wing, including 614, were called for support. Two 614 crews went to their aid, arriving at Souk El Arbaa. They were Bisleys BA869 of F/O Georges with Observer F/S John Taylor and WOP/AG Sgt William Sorbie, and BA734 of F/O Bill Irving with Observer F/O Leonard Quevatrc and WOP/AG Sgt Limoges. Two further 614 Bisleys, with mixed crews from other squadrons of the Wing, would join them from Blida.

They were Bisley BA800 flown by Flt Lt A Breakey of 18 Squadron and BA825 flown by P/O GW Sims with his crew from both 18 and 13 Squadrons.[23]

A promised fighter escort of Spitfires was cancelled. Another 18 Squadron pilot (who would join 614 Squadron some months later), Ted Holloway, remembered Malcolm's reaction when told of this. 'Bull****,' he'd replied.[24] They would instead make a 'sweep' of the target. Malcolm walked to his aircraft with Holloway, dejected. They mused that they would be 'in for a rough time'.[25]

They took off at 15.15 and were in trouble almost immediately. F/O Armstrong, a 13 Squadron pilot, burst a tail wheel when taxiing and got left behind.

Just 20 minutes into the flight, the 614 Squadron Bisley BA825, flown by P/O Sims, had engine trouble and crash-landed 15 miles east of Souk El Arbaa. Sims and his two crewmates survived.[26]

Malcolm now had just nine Bisleys, all flying in three 'boxes' of three. The fighter sweep had taken off ten minutes before them – so they were on their own. The bombers kept close in tight formation – their only protection against enemy fighters.

Bill Irving, from 614 Squadron, was flying number three in Malcolm's formation. From Cwmbran, he was originally an Army officer who survived the Dunkirk evacuation by commandeering a lifeboat and had rowed back to England before being picked up. He'd sworn then that the only way he would ever return was by air, and had transferred to the RAF and trained to be a pilot. He joined the squadron formed just 15 miles from his home town in 1941 while it was based at Macmerry.

The three formations gently rose and fell on the sharp updrafts from the mountains and Ted Holloway recalled, 'We bypassed the target, taking advantage of hill cover to hope for some element of surprise. We approached from east to west for a bombing run at about 1,000 feet.'

Gordon Wallace flicked the bomb master switch and buttoned his face mask. They'd be over the target in less than two minutes. He quickly glanced around from the perspex-glazed nose of his Bisley. All seemed clear. Suddenly Henry Parsloe, his gunner, shouted. Aircraft were above them. Wallace looked up and saw more fighters than he'd ever seen before. He assumed that they were the Spitfires sent to sweep over the airfield. But these were no Spitfires. An Me109 flashed low across them from ahead. Parsloe called out that they were under attack from a swarm of them from all directions. Another Bisley passed close below Wallace's aiming panel, its starboard engine on fire and streaming a plume of dense black smoke.[27]

Ted Holloway recalled they estimated there were 60 German fighters. Their box broke up and his aircraft came under continuous attack. 'We fought them off, turning into each attack. We were soon hit and the starboard engine went on fire.'[28]

Meanwhile, Bill Irving's Bisley came under fire from a different direction. One engine was knocked out and he left Malcolm's formation, weaving to evade three 109s. The gunner, Sgt Limoges, hit one before his guns jammed, but beat off another two when he began firing again. Flying low along the valley floor to avoid the pursuing fighters, he crash-landed at Souk el Khemis where he, F/O Quevatre and Sgt Limoges, clambered unharmed out of the mangled bomber, and walked to the safety of Allied lines.

The sparse entry in Irving's log book gives a glimpse into those desperate minutes but also his skill in nursing his stricken bomber out of danger and saving the lives of his crew.

'Attacked by three 109s. Starboard aileron 3/4 shot away and cannon shell in port motor on first attack. Rudder and elevator trimming tabs shot away and port engine stopped after five minutes. Flew for twelve minutes afterwards on one engine. Complete attack lasted for 18 minutes. Crash-landed at Souk el Khemis 20 miles from base. Gunner had one confirmed, one probable and one damaged.'[29]

So vast were the odds against the Bisleys, the Germans were almost having to queue up to fire at them. Having shot one bomber out of the sky, Oberleutnant Julius Meimburg was one of those flying the 109s that afternoon. He recalled, 'Several Bf109s at this time are in quite a crowd behind them. I then shoot down a third which goes down burning and crashes. I can only fire a little at the fourth. I attack as all my ammunition is then gone. I then attacked one on the left and, as I am flying in a curve, I can see five already shot down.'[30]

Two more Bisleys of 18 Squadron were hit and crashed behind enemy lines near Chouigui with no survivors. Other Bisley gunners hit German fighters. One 18 Squadron aircraft claimed three, another claimed two. But it was hopeless. The sky was full of them and they overwhelmed the bomber crews.

Holloway recalled, 'The engine fire had by this time spread along the wing – flames were coming into the cabin and Wallace began to get burned.' Barely able to control the stricken bomber, he belly-landed on the desert floor. The three clambered out and got clear just before its unreleased bombs started to explode. They were rescued by a British Army unit nearby.

Another Bisley from 614 would not be as fortunate as Irving's or Holloway's. A crashed aircraft had been seen by the soldiers and Holloway was taken to identify the dead pilot nearby. 'It was F/O Georges,' Ted recalled. 'His parachute had not fully opened – not a pretty sight.'[31]

The soldiers had seen Georges' aircraft being attacked by six 109s. The cockpit was on fire and Georges had attempted to bail out but his parachute failed to open. Both WOP/AG, Sgt William Norton Sorbie, and Observer, F/Sgt John Taylor, were also dead.[32]

The battle had lasted just minutes but nine Bisleys had been cut from the sky and 18 aircrew had been killed.

Hugh Malcolm was awarded a posthumous Victoria Cross the following April for the gallantry he showed in leading this dangerous, but doomed raid and two previous missions. A chain of rest and recreation centres in the Mediterranean theatre were named 'Malcolm Clubs' in his honour.

The action would force a decision on how the Wing operated and was another damning verdict of the Bisley's capabilities.

Peter Le Cheminant recalled that it was a watershed

moment in the debate on whether the Bisley should only be used at night or if it could operate in daylight.[33]

Charles Bonner, leading the party arriving from Setif, recalled how when they arrived at Canrobert the next day, they were stunned by news of the raid. The squadron diary noted, 'The CO and party hear news that P/O Georges and his crew were killed in a daylight attack with 18 Squadron, who were almost entirely wiped out. F/O Irving crash-landed and walked back to safety.' Strangely, just months later, Bonner would become 614 CO, so too would Irving in 1946 when the squadron returned to Wales after the war's end.

It would be the most disastrous day in the squadron's history thus far. While reports came through of the death of P/O Georges, Sgt Sorbie and F/Sgt Taylor, there was also news that day of further losses with confirmation that F/O Thomas Young and W/O1 George Campbell had been killed, and WOP/AG Sgt A Walsh was injured after their Bisley had failed to return after a raid on Tunis Aerodrome two days earlier.

The 326 Wing boss, Group Capt Sinclair, decided to respond quickly and planned another daylight attack in the same area. Peter Le Cheminant thought that this was to lift morale, dented as a result of the raid and 'perhaps partly motivated by a desire to stick two fingers up at the Germans.' Six 614 Bisleys were ordered to again attack the airfield near Chouigui, but this time there was an escort of 24 Spitfires. He recalled they successfully bombed the target, but at 6,000 feet, as opposed to 1,000 feet, the height Malcom's raid had flown at 'to a degree' he recalled, 'it was felt that honour had been satisfied'.[34]

Amid the losses, there would be recognition for one airman's bravery. Sgt John Hamer Vickers was WOP/AG of a Bisley which attacked Bizerta docks, also on 4 December. His aircraft was hit by flak and Vickers was wounded in the foot while his parachute and dinghy were set on fire. Despite his injuries he left his turret, which he operated by hand as the hydraulics had failed, and threw the parachute and dinghy out of the aircraft, putting out the fire and enabling the Bisley to be brought back to base.[35]

Following the disastrous Malcolm raid, the squadron and the rest of the Wing concentrated on night attacks, but were hampered by atrocious weather for much of December. Torrential rain turned airfields like Canrobert into quagmires. Metal tracking, used in the hope that the sodden airfields could operate, simply sank into the mud.[36]

It was a frustrating time with sorties constantly cancelled because of the conditions. 'Canrobert Airfield unsuitable for operations – ground too soft'; 'Mud and weather too bad to take off', the diary told. Both 18 and 614 Squadrons used the remote airfield, which lay some 3,000 feet above sea level on a plain in the Aurès mountains. Although 614 hadn't been in its new home for a month, the squadron was also again looking to move. Canrobert, while nearer the frontline than Blida, was still over 200 miles from Tunis.[37]

They still had a much longer and more difficult flight to reach their objectives than the enemy, who occasionally attacked the airfield. Following a raid on 16 December, 'all available bodies' were engaged in digging slit trenches.[38]

Some lucky ones found shelter in a small French Air Force hangar at the airfield's edge. But most were still living under canvas and many became ill in the cold and damp. In an effort to raise morale amid the filthy weather and the stop/on operations, the CO instructed the men around the 20th to take plenty of exercise. 'Mountain climbing very popular,' noted the diary. Two days before Christmas all squadrons organised 'bathing for men at hot Turkish baths at Aïn Beïda'. Christmas Day saw the men enjoy a 'splendid Christmas dinner of fresh pork' and a gift of '£50 worth of cigarettes, etc.' from south Wales, donated by the Glamorgan Territorial Association to give the men something extra.

But on Boxing Day the diary recorded, 'Christmas festivities over,' and 614 prepared for operations that night. 'All aircrews and groundcrews working on aerodrome. Target tonight – roads to Tunis and Pont Du Fahs – briefing at 16.30 hours.'

The morale-boosting festive treats did little to alter the view of Canrobert as a god-forsaken quagmire. Bonner recalled a song which summed up their feelings on the base. Sung to the tune 'Oh my darling Clementine', it made their feelings clear:

In the desert, by a mountain,
F***ing miles from anywhere,
There's an airfield, bloody airfield,
And they call it Can-rob-air![39]

Such are the vagaries of war, while much of the conditions at Canrobert were dire, there were moments of relief, like the Christmas dinner. Peter Le Cheminant recalled being able to fly to Algiers where he and his crew enjoyed a five-course lunch and he bought a bottle of Chanel No. 5 perfume for his wife.[40] LAC Andrew Turnbull, meanwhile, resorted to keeping hens whose eggs helped supplement the endless bully beef.

Early January saw the squadron prepare an alternative base at Oulmene. Slit trenches and latrines were dug, and construction of an underground operations room was begun. This preparation would be vital for the momentous month which lay ahead. But there were soon losses. Following a raid on Tunis docks, Bisley BA731 failed to return and the crew, Pilot F/O Martin Stanley Morris, W/O II Raymond Marsden, RCAF and Sgt Findlay London Mills were all killed.

There was some relief when both weather and moonlight allowed more operations. The bombs of 614's Bisleys regularly fell on transport in southern Tunisia, around Sousse, Sfax, Kairouan and Enfidaville. The road and rail crossing at Pont Du Fahs was also a regular target, and all would prevent the movement of Axis forces.

The CO, Wg Cdr Sutton, was showered with awards in

the new year. On 1 January he was awarded the OBE for 'his outstanding powers of leadership and by his personal example producing a magnificent spirit in his squadron'. Citing the successful part he played in the Thousand-Bomber and Dieppe raids, it said his 'determination and skill have earned him the complete support and confidence of his crews'. On 26 February, he was awarded the DFC for 'the great success achieved by his unit, directly attributable to his personal leadership'.

As the number of missions crept up, so the squadron lost more aircraft. None were due to enemy action, but to burst tyres on take-off. The rough, stony airfields took their toll as the Bisleys strained to lift off at full power and with a full bomb-load. On 16 January, one of F/O Hanbury's engines cut out on take-off and he crash-landed his Bisley three miles from base. One of Le Cheminant's tyres burst on take-off and his undercarriage collapsed.

In the wake of the disastrous raid led by Wg Cdr Malcolm and constant attrition, new aircraft and crews were ferried out from the UK. Following the raid in December, 18 Squadron had all but ceased to exist as a fighting unit and would later re-equip with the far-superior Boston light bomber.

Another casualty was Sub Lt Francis Cleak, of the Royal Navy. Like Gordon Wallace, he was an observer seconded from the Fleet Air Arm. He died from injuries sustained in a motorbike crash and was buried in Constantine with military honours given by French Army.

Towards the end of the month the diary noted that, 'Work at Oulmene proceeding very satisfactorily'. The tempo of offensive operations had stepped up. But they could have no inkling of what February would bring.

By the end of January 1943, Allied forces, which had been slowly advancing following the 'Operation Torch' landings and supported by 614, would face a new opponent. Rommel, retreating after El Alamein, reached the south of Tunisia.

Taking stock of the situation, the Axis commander at once decided to safeguard his communications – and line of retreat – before giving battle to Montgomery.

By 7 February, 614 had completed their move to Oulmene. There had also been sporadic operations attacking German positions in the first half of the month, but bad weather had again meant flying was limited.

After the 'Race for Tunis' in November and December, the Allies had become bogged down in the mud, and Eisenhower waited for the weather to improve before moving on to Tunisia. Mainly American and French forces were scattered thinly in the hills separating Axis-occupied Tunisia and Allied-held Algeria.

The threat might have seemed distant to 614 at that time, but the danger was mounting. By 14 February, Rommel descended on the critical town of Sidi Bouzid. The thinly-scattered Americans were overwhelmed.

Roads leading to the town were strewn with knocked-out US tanks, half-tracks and artillery. Two days later the Americans withdrew to Sbeitla. The Germans advanced to occupy Gafsa against little opposition. Roads were jammed with fleeing troops and vehicles. US aircraft dropped messages to surrounded American battalions, telling them to escape.

No. 614 too was set to take on an air-drop. 'Emergency preparations for immediate dropping of food containers to our front-line troops cut off,' read a diary entry on 16 February.

If the advancing Germans could break through at the Kasserine Pass and through the mountains into Algeria, the way was open for them to spill out westwards.

The squadron diary on 17 February betrayed little sign of the peril that was looming. That was, until just before midnight, when the squadron received a desperate message from Group Captain Sinclair at Wing HQ, 'Huns break through at KASSERINE – emergency maximum effort tonight.' The line reads like a newspaper headline rather than a sober diary entry and conveys the dire situation the squadron found itself in. Six aircraft took off to attack targets in the area, but they could see little through the cloud.[41]

By 18 February the two Wings of the offensive had met; Rommel's battlegroup coming from the south and General Arnim's forces, in north-west Tunisia, linking up at Kasserine and threatening to break through the central mountains of Tunisia and roll the Allies back through Algeria. The battle for control of this vital mountain pass was joined.

The prospects did not look good. Air power so far had not been able to stop the Axis advance, but 614 would be part of the effort to do just that. The diary read, 'Special orders received to make aircraft serviceable for operations. Emergency flap re KASSERINE PASS maximum effort for tonight. Nine aircraft and crews prepared.'

Le Cheminant recalled how the 10th Panzer Division was reported to be advancing rapidly and was only 40 miles from Tebessa over the border. By 19 February, the Germans continued to advance and approach the Wing's bases. Le Cheminant remembered, 'the possibility of the Germans breaking through and reaching Canrobert and Oulmene was beginning to be taken seriously. There was a lot of rumour flying about such as that the Huns were only six miles from Tebessa and, consequently, a certain amount of panic in the camp.'[42]

By 20 February the Germans were still pushing the Allies back through the pass – clogged with mud in the cold and wet. That day, at Oulmene, Sinclair called a meeting. He pulled no punches in warning of how desperate the situation was. There was a real risk they could soon be overrun by the advancing Germans. They had to do all they could to strike them from the air but also be fully prepared to fight them hand-to-hand if they reached the airfield.

He spoke 'calmly and candidly' and told them that the

situation was so serious that they should 'be prepared to engage the enemy to the last aircraft'.[43]

Le Cheminant admired Sinclair and saw him as one who led from the front. 'Sinc, true to form, responded in swashbuckling terms and announced that, if there was a breakthrough, the aircrew would take to the mountains overlooking the village and operate as guerrillas.' Although a brave flyer, Le Cheminant baulked at this and recalled his diary entry read simply, 'Not my cup of tea at all!'

So real was the threat from the Panzers, Sinclair had the anti-aircraft guns that ringed the airfield moved up into the hills surrounding Oulmene. He instructed gun crews to lower their gun barrels so that they could act as field guns if they broke through.[44]

By dawn on 21 February, although Rommel's forces were on both sides of the Kasserine Pass, elsewhere, the Axis position was not as good. British and American delaying actions had already stopped the right prong of its advance.[45]

That night, nine 614 aircraft were readied for the night's raids and all crews were operating under exceptionally bad weather and 'full praise was due to squadron members'.

The improvement in co-operation between air and ground forces was felt at Oulmene. On 21 February, the Wing was joined by an Army liaison officer, Major Wordsworth, who would attend all operational briefings. This was seen as 'very satisfactory as crews get a picture of the battle before setting out'.[46]

Losses of aircraft and crews continued, with the Wing losing three Bisleys and one WOP/AG killed trying to bail out. Reinforcements continued to arrive from the UK[47] and Sinclair recalled they were lucky to get these additional Bisleys and crews. 'We continued flying by night but the weather got worse and enemy defences stronger and my casualties mounted. Things became so bad that they milked the Blenheim's OTU and sent their aircraft out to bolster my numbers.'[48]

By 23 February, the news from the front was improving. For the Wing, Sinclair noted that the situation was 'not so critical as it was'.[49] The day after saw Le Cheminant briefed to bomb enemy concentrations further south, at Gafsa, with incendiaries which 'left it burning satisfactorily' he remembered.[50]

On 25 February the call from Sinclair was, 'Maximum effort for tonight', with a raid in preparation. The attack would be in support of the Army in the Kasserine Pass and Sbeitla. That action would see one 614 pilot, Ian MacLaren, decorated for his bravery in pressing home his attack on enemy positions in the pass.[51]

Like all the Bisleys in the Wing, they were very heavily laden with fuel and bombs. The risk of burst tyres and accidents on take-off were ever-present. Tragedy struck when a Bisley from 13 Squadron crashed into a lorry and killed the two 614 airmen inside, LACs George Henry Hawken and Leo Woodburn. The crew escaped but the number of sorties that night was restricted due to bombs exploding on the still-burning aircraft.

Another Bisley crashed when it attempted to take off in the teeth of a storm. WOP/AG in MacLaren's aircraft was Paddy Stapleton, who recalled lifting off that night with great difficulty and trying to find the Kasserine Pass. The weather was filthy and low cloud obscured the target. They followed the railway line and, as they sped down the valley, looked for the gap in the hills.

Flying low was often the key to finding the target, he remembered, but this exposed them to anti-aircraft fire and they often narrowly escaped taking hits. 'The sky would be thick with flak, and searchlights would blaze to light up their targets.' They flew alone, and so there was no covering fire from other Bisleys.

The flak was merciless. 'When we got there the enemy was very well dug in and gave us great opposition.' He recalled ruefully, 'The Afrika Korps had plenty of shit coming for us.' But Paddy gave as good as he got and blazed away from his turret, firing his twin Browning .303 machine-guns. Making more than one pass, they hit transport on the road from Kasserine to Tebessa and bombed a train in Kasserine sidings.

When they returned to Oulmene, a groundcrewman asked a bemused Stapleton, 'What on earth have you been doing in the back there, Paddy? Tearing up sheets of paper?' He found that the parachute on his back had been shredded as the flak burst around them and shrapnel sliced through their aircraft. He recalled they counted 103 bullet holes in the machine.

At debrief they found that Army units on the front line had seen the effect of their action. 'They'd reported that we were persistent in the attack. We apparently stemmed the gap and saved the day.'[52]

For this operation Sqn Ldr MacLaren was awarded the DFC for the 'real courage and determination' he showed in attacking enemy transport and troops.[53] Also awarded the DFC were F/Os Robert Baelz and Norman Dumont for their determined and courageous attacks on the Kasserine Pass in the face of very adverse weather. Montgomery and Air Marshal Tedder signalled commendations to 614 Squadron after their strikes left roads littered with burnt-out shells of enemy vehicles.

In the wake of raids like MacLaren's, Axis forces withdrew. The offensive had been faltering for the past few days and now it became a rout. The squadron had played a key role in what was a desperate fight. It had shown the stuff of which it was made and developed its tactics to succeed. That 614 had been able to play its part in the battle of the Kasserine Pass was, in part, due to improved co-ordination of US and British air power and between air and ground forces.

'Operation Torch' was the first combined operation between the Americans and British in the Second World War. Inevitably, there were problems around co-ordinating two huge and differing forces, especially so in the command and control of Allied air forces. The Allies approved a plan to unify the different air forces at the Casablanca Conference in mid-January 1943. This led to the creation of Mediterranean Air Command under Air Marshal Tedder.

This reorganization was applied at the most critical moment and was introduced on 18 February, just as the Axis forces were crushing resistance on the Kasserine Pass.

The permanent placing of an Army liaison officer (ALO) at Wing briefings was an indication of how that thinking was quickly pushed down the chain meaning, by February 1943, Army intelligence was being put in front of 614 crews on a daily basis and questions could be asked and information shared. This was to prove vital and helped turn the tide.

Following the near-disaster at Kasserine, Eisenhower brought the First and Eighth Armies under a new headquarters and the leadership of General Alexander.

Now, the 'co-ordinated land, sea and air power of the Allies would draw a net round the Axis forces in Tunisia by 30 April, to meet the timetable set at the Casablanca Conference to allow Sicily to be invaded during the favourable weather of August.'[54]

Like the Allies, the Axis also formed a joint command for their armies. Rommel was given the command of the new Army Group Africa on 23 February. He had pulled back from Kasserine with the intention of facing Montgomery once more, who was approaching southern Tunisia.

As the Desert Fox pondered his next move, 614 grew ever-more discontented with their aircraft. Ever since they were equipped with the Bisley in the autumn of 1942, many in 614 and in the Wing had been unhappy with them. Andrew Turnbull recalled of the other units (of 326 Wing), 'Some say they're getting Bostons, others say Mitchells, but whatever they get it's a bit hard on us getting these old Bisleys. They're out of date and they should have been obsolete [*sic*] years ago. They are definitely hopeless now.'

Meanwhile, the underpowered Bisley continued to crash, killing more and more aircrew of the Wing. New aircraft and crews were sent from Britain, but morale suffered. Le Cheminant recalled at this time that they were losing aircraft 'at a ridiculous rate – most of them in no way due to enemy action'.

He estimated they had lost 30 aircraft by the beginning of December, with another 25 during that month because, by 31 December, they had 16 aircraft remaining out of the original 72. On 1 January six new aircraft arrived, bringing the Wing up to 22, and a day later a further six, but on 6 January Gp Cpt Sinclair reported to headquarters that he had only twelve serviceable aircraft. On 13 January Le Cheminant did an analysis of the fate of the 18 aircraft with which 614 had left England and found that only four of the originals remained on squadron strength; ten had been written off and four were given away to plug gaps elsewhere in the Wing.

Having to operate these heavy, under-powered aircraft at altitude was affecting morale. He wrote in his diary, 'most people getting rather jittery about flying these bloody aircraft – engine cuts getting more and more frequent'.

In the last half of January there were two fatal crashes. The morning after the second one a crew of 18 Squadron came to him as controller and asked to be left off the battle order as they were very shaken after the previous night's crashes.[55]

The Wing's medical officer reported this too. 'Aircrew morale at one stage was somewhat lowered, entirely due to the frequent accidents which occurred with the Bisleys, for which the crews were not to blame. Their defensive armament was poor, and aircraft would not maintain height on one engine.'[56]

A lament was penned by Flt Lt Frank Desmond Win, a New Zealand pilot with the squadron, which expressed their feelings on their unreliable aircraft. Called 'The Bisley Song', it was sung to the tune of 'Marching Through Georgia':

> Here we are, the Bisley boys of 614,
> Always in some trouble but we still come back for more,
> May not see many Jerries but we still have prangs galore,
> 'Cause we're still flying the Bisleys
>
> Chorus:
> 'This kite, this kite, is no bloody good,
> It does the things no decent aircraft should,
> When we get back to civvy street we'll be so ****ing rude,
> And tell them where to stuff their Bisleys.'

Charles Bonner stoically flew on with his crew. The feelings he recalled having before an operation would have been familiar to many on 614. 'Our thoughts always the same. Impossible to sleep beforehand. I wonder what's happening at home? What will it be like tonight? Will a tyre burst? Will the engines cut? I hope that there is someone up there watching over me!'

Some squadrons in the Wing did begin to get Bostons – but not 614, although some from the squadron got to fly them.

By 24 February, although he did not know it at the time, it was to be Le Cheminant's last sortie in the cursed Bisley and with 614. He had flown 17 operational sorties for a total of 109 hours, and had written off two of them. He was chosen to be the next CO of 114 Squadron which was to re-equip with Bostons, and he was sent to England to pick one up.

Reflecting many years later, he still was bitter about how they had had to fly the 'sub-standard' Bisley which resulted in 'the operations of the Wing can only be seen as a minor disaster'.

He continued, 'Now, after 57 years, I'm prepared to concede that it may well have been necessary to have a British light bomber in production to tide some squadrons over until enough Bostons and Mitchells became available. What I do not concede is that it should have been the Bisley.'[57]

By 20 March, Sinclair had brought the British and American medium bomber units together to create a new formation. Now promoted to Air Commodore, he took the helm of the Mediterranean Allied Tactical Bomber Force, with the American Maj Gen Fred Terrel as his deputy. The Bisleys of 326 Wing would now be joined by the American A-20s (or Bostons to the British).

Sinclair was still keen to re-equip the British squadrons with the Boston. He recalled, 'I gave one of the Bostons to each of the ex-bomber squadrons [18 and 114] and told them to convert all their pilots as quickly as possible.' But 614 and 13 would have to soldier on with the unloved Bisley. Here, the different backgrounds of the four squadrons in the Wing came to the fore. Both 13 and 614 Squadrons were army co-operation units and Sinclair perceived them to have different levels of expertise. When deciding who would have the Bostons, priority was given to the bomber squadrons.[58]

Meanwhile, on the ground a final reckoning was building with Montgomery's Eighth Army massing to the south near the Tunisia-Libya border. The Axis powers risked being squeezed between them and the combined US-British forces of the First Army, backed by the airpower of 326 Wing and more approaching from the west.

On 6 March, the Germans and Italians launched 'Operation Capri', an attack southward towards the northernmost British strongpoint. Its failure convinced Rommel that the Axis cause here was finished and, on 9 March, he travelled to Italy and later to Berlin to convince Hitler that North Africa should be abandoned. The Führer disagreed and had Rommel placed on sick leave. He never returned to North Africa and Montgomery would never have a final showdown with the Desert Fox. Arnim became commander of Army Group Africa.

The squadron's operations over the next month or so

would be in support of the Eighth Army advancing from the south and the First Army from the west.

On 17 March, the US 1st Infantry Division took the town of Gafsa, and the next day the oasis of El Guettar fell with little opposition. In the south the night of 19–20 March saw the Battle of the Mareth Line joined as Montgomery launched 'Operation Pugilist' against Axis defences. In the week following the taking of El Guettar, the US forces slowly moved forward to take the rest of the interior plains and set up lines across the entire Eastern Dorsal mountains.[59]

As 'Operation Pugilist' faltered, Montgomery tried to outflank the Mareth Line with the New Zealand Corps pushing north-west for the Tebaga mountain gap.[60] An attack on 21 March by twelve Bisleys from 614 on a landing ground there supported their advance.[61]

Charles Bonner was part of the raid and, having avoided heavy flak, they flew south-east a little to 'see how the Eigth Army assault was going'. He recalled, 'From 8,000 feet, it was amazing to see the volume of artillery fire being directed against the German-held Mareth Line', seeing 'miles and miles of rippling red fire'.

Two days later, another twelve-strong sortie left to attack roads around the Sfax to Sousse and Kairouan areas.

Earlier in the day, Air Cdre Sinclair visited Oulmene and placed 614's Flt Lt Wallace in charge of building a 2,000-foot-long runway with taxiing tracks. The need for a proper runway was underscored many times with the continued loss of Bisleys on the rough airfields. The improvements to Oulmene continued with the completion of the underground operations room the next day, which was pressed into service immediately for that night's raid.

Briefed to attack enemy troop concentrations and roads in the Sfax to Sousse and Kairouan areas 150 miles north of the ground battle, Charles Bonner recalled seeing a column of smoke rising from near Mezzouna. Believing it to be enemy positions and vehicles, he dived to attack. Suddenly he came under heavy fire. 'Bang! Streams of ack-ack came at us, with many sharp intakes of breath, and a very loud "Christ!" We were hit and started to dive to starboard. The damaged Bisley was slipping out of control and the ground was looming fast. By giving the aircraft full port aileron, and heavy port rudder, I was able to straighten her out and, with some hard yanking back on the control column, I was able to stop the twisting dive, and level out.' The flak was still coming, but was now to their rear. 'In order to gain height we literally threw our bomb load at the cross-roads near Mezzouna,' he recalled. 'It was impossible for our navigator to be more accurate, as we were going through the air like a flying crab, and the bomb-sight was ineffective.' To keep control of the damaged bomber he flew them home with his left arm stuck through the left-hand loop of the control column head, and his left-hand clinging to the underside of the instrument panel in front.

When they landed back at Oulmene, they found a shell had gone through the starboard wing tip, causing the upper metal skin to peel back from the ribs and lie back along the aileron, severely limiting any movement.

Another song penned by Flt Lt Frank Win, 'Tales of Kairouan', sung to the tune of 'Oh Christmas Tree', told of the dangers they faced on these missions:

I'll tell you tales of German flak,
At Kairouan, at Kairouan,
It comes up thick, it comes up black,
At Kairouan, at Kairouan,
You've only got a Bisley 5, so if you want to stay alive,
Press on the stick and dive and dive,
And get away from Kairouan.

Despite the flak, all aircraft returned and Sinclair signalled to compliment the squadron on a 'bloody good show'. There were congratulations too from the US II Corps who were advancing from the west hoping to link up with the New Zealanders advancing from the south. The information secured by Wg Cdr Sutton's recce of the front line was reported to have made 'a big difference to the battle picture' that morning.[62]

As the campaign progressed, so the Allies' use of air power improved. The Wing flew target-marking missions loaded with incendiary bombs and flares. Flight commanders of the four squadrons would fly out at dusk to mark where the enemy tanks and fuel supplies were concentrated. Further waves would then follow with incendiary bombs to set the area alight; then, with high explosives.[63] This would presage a future incarnation of 614 squadron as a Pathfinder unit.

As the Eighth Army pushed north and the First Army east, the Axis were in danger of having thousands of men and their equipment encircled and, between 28 March and the first week of April, they withdrew towards the north-east.

By 7 April the Americans of II Corps raced down the El Guettar to Gabès road, where it met advanced troops of the Eighth Army. The forces of the eastern and western deserts had joined.[64]

At 11.20 that day, a special message came through to 614 from the TBF (Tactical Bomber Force), 'The Hun is on the run and it is reflected that a maximum effort will be required tonight.' The squadron was 'on its toes and eager to beat all other previous wing monthly sortie totals'.[65]

The pursuit covered 140 miles to the north of Wadi Akarit. The Axis troops fell back to defensible positions towards Enfidaville, 25 miles south of Cape Bon. The net was closing.

On 9 April, Sinclair expressed his pleasure to Nos 13 and 614 Squadrons at the past work of the Bisleys, and stressed the importance of their work in the coming battle. One of the main jobs now, Sinclair told them, would be reconnaissance off the coast, 'watching for disembarkation of the Hun'.

Bisleys of 614 would carry out the general instructions

of Admiral Cunningham, Eisenhower's Naval Task Force Commander, to 'sink, burn, capture and destroy' enemy shipping. 'Let nothing pass,' he declared.[66]

Because they flew at night, 614 had to wait for the moon to reappear, so operations paused in the first week of April. Aircraft were made ready for the next phase of the offensive. The squadron diary noted, 'Serviceability must be maximised for the coming month,' and paid tribute to the groundcrews who were, it said 'doing fine work on aircraft maintenance'.

There was time for the men to enjoy some diversion. Football matches were arranged in the evenings between flights or sections. All ranks took part with 'great enthusiasm amongst the men', the diary noted.

As the Eighth Army advanced north towards Tunis, the town of Enfidaville stood in the way. By 19 April, the battle to take the town began when the Eighth Army attacked it, meeting strong resistance. In the days preceding this battle, 614 helped soften up the town's defences, and the rest of the month would see the squadron operate at an unprecedented tempo. On 10 April, the message came through from the TBF, 'Maximum effort required for tonight' and 614 boss, Wg Cdr Sutton, set a target for 60 sorties for the night.

With 13 Squadron, 614 responded with 59 aircraft taking off and successfully bombing Enfidaville, setting a record for a single night's effort for the Wing.

This effort was increased in the following days. Practically every night the Bisleys took to the air, pounding Axis positions at Enfidaville or enemy airbases nearby. Targets such as aerodromes at Sainte-Marie du Zit, Fochville (near Tunis, now called Ben Arous), La Sebala, La Marsa (near Tunis) and Sidi Ahmed were pounded, and the enemy's capacity was gradually eroded. The docks at Tunis were also hit.

The attack on Sidi Ahmed Airfield on the night of 19–20 April saw the two squadrons attack with 30 aircraft bombing the target in an eight-minute window to swamp defences. This method was used repeatedly until the end of the campaign in North Africa.[67]

Charles Bonner remembered how effectively the squadron was operating. In one attack on Bizerta Airfield in late April, he and Wg Cdr Sutton circled the airfield, waiting for their moment before dropping incendiaries across two intersecting runways. 'It was uncanny… we flew around the airfield, with the runways reflected in the moonlight, not a gun was fired at us, indeed there was not a movement anywhere.' The twelve other Bisleys from the squadron orbited, waiting to drop their bombs on targets marked by them. Seconds after the incendiaries and flares were released, flak began to burst all around. The bombers dived through the exploding flak and let loose their loads. Miraculously, none were hit and all made it home. 'It was a very noisy party, but we all landed safely, as we had all been briefed to fly at specific heights and, after dropping our bombs, to leave the target area, reducing the risk of collision to a minimum.'

On 20 April, the Eighth Army captured Enfidaville in the last major battle they would fight in North Africa. The Axis was not yet defeated, but their days in Africa were numbered. For 614 Squadron, the month had seen their highest tempo of action of the war. By 25 April, they had flown 155 missions totalling 433 hours, and dropped 268,200lbs of bombs. The diary recounted, 'This far outclasses any previous records. The air and groundcrews of both squadrons have done a fine job of work.'

The final phase of the Allies push in Tunisia, 'Operation Strike', began on 5 May and 614 Bisleys lifted off, just hours before, to attack enemy transportation. The 17 aircraft dropped 28,800lbs of bombs on vehicles between Tebourba, Massicault, Bir Meherga and Cheylus.

The following night the squadron again prepared for maximum effort, with 18 Bisleys dropping 30,890lbs of high explosive (HE) and 1,080lbs of incendiaries on airfields at

El Aouina, La Sabela and Ariana. Wg Cdr Sutton, Sqn Ldr Bonner and Sqn Ldr McLaren acted as pathfinders for this mission and the diary noted, 'All aircraft returned safely, and although the night was very dark, fires were started on all three aerodromes.'

Paddy Stapleton remembers these missions. They faced 'heavy opposition' from searchlights and flak. He recalled facing especially heavy fire during this attack on El Aouina Airfield, with shrapnel splintering the nose panel of the aircraft. He admitted teasing their navigator, Denis Ager, saying that he had developed a twitch following the flak on this operation.[68]

On 7 May, at 20.00, the fall of Tunis and Bizerta was announced; 'party in the Mess tonight' the diary recorded.

The enemy was now cornered on the Cap Bon peninsula, south-west of the capital. The pace of the advance on the ground was quickening still, with daily briefings from the ALO telling just how close the air support was needed for the troops on the ground. The squadron diary reflected this ever-changing picture. On 8 May, it noted, 'Bomb line continues to advance,' before telling of yet another maximum effort which was needed for the night. Flares were carried by all aircraft and 32,150lbs of HE were dropped on roads around the peninsula. On 9 May, at the usual 18.00 briefing, crews were given their targets – again, beaches and coasts on the peninsula. In all, 17 sorties were flown – dropping 28,800lbs of HE on the enemy.[69]

On the night of 10 May, the targets were again roads and beaches on the peninsula. However, crews returning from their first operation learned that the enemy line in this sector had practically collapsed, and their second target that night was reduced to the remaining corner of the peninsula where resistance continued. Both 614 and 13 Squadrons flew 66 sorties, dropping 112,460lbs of bombs and 33 aircraft dropped flares to mark targets. It was to be 614's final act of the North African war. The next day the diary said simply, 'The capitulation of the bulk of enemy forces in Tunisia is announced. No target for tonight.' By 12 May, German and Italian forces had surrendered. A day later, Air Cdre Sinclair sent congratulations to Wg Cdr Sutton for the squadron's efforts.

On 15 May, the whole squadron turned out for a CO's parade. The boss, W/C Sutton, revealed the extent of 614's effort for the past six months. He praised the 'highly successful' squadron and the high morale of air and groundcrews. Later in the afternoon, the officers were entertained by the station orchestra. In the following days there would be swimming at nearby Foum de Gueiss.[70]

But the Allies didn't celebrate for long. Their focus was now on Sicily, which was invaded on 9 July 1943, only two months after the Axis collapse in Tunisia.

It was a momentous time. Allied troops would land on the continent after having been driven out three years previously. The campaign in Italy would be a long and costly one in which 614 Squadron would play a part but would see many of its men fall. But, for now, they wouldn't be going to Italy. Number 614 Squadron would play a part in those landings, but having started life as an Army Co-operation unit, then intruder, then light bomber, they would now have to turn their hand to yet another role to help protect the convoys of ships that would bring men and machines to Sicily for the invasion of Italy. They would patrol the Mediterranean Sea, protecting convoys.

CHAPTER 5

Coastal Patrol (1943–1944)

FOLLOWING THE AXIS surrender, the squadron would soon leave their home at Oulmene and go back hundreds of miles west to the vicinity of Algiers.

They had all arrived at Tafaraoui by 26 May and, on that day, as aircrew paraded in the hangar before Wg Cdr Sutton, he announced that the squadron would henceforth operate for Coastal Command.[1]

At Tafaraoui, 614 joined another AAF squadron, No. 500 (County of Kent) who were already experienced in an anti-submarine role, having been there since the beginning of 'Operation Torch'. They would later donate some of their experienced crew to 614 to give them a leg-up as they learned their new role.

The squadron became part of the North-West African Coastal Air Force (NACAF) under Air Vice-Marshal Sir Hugh Lloyd. Still flying the Bisley, 614 were joined by their comrades from 326 Wing, No. 13 Squadron. Alongside them were a variety of units flying an assortment of planes, including the Hudson maritime patrol bombers and also Wellingtons and Halifaxes adapted for a maritime role.

The first few days of June saw aircrew learning their new roles. The rest of the squadron, having arrived by road from

Oulmene 380 miles away, were allowed some recuperation at Les Andalouses, a beach resort just up the coast.

The Bisleys were soon modified. Out went the bomb-sight, giving the navigator an unimpeded view through the glazed nose, and a chute was installed for dropping flares at sea in the forward escape hatch.[2]

One of the aircrew who transferred to 614 from 500 Squadron was WOP/AG Rex Thompson. For them it wasn't a popular or welcome move. The New Zealander recalled, 'At the stroke of the pen some bright spark along the chain of command decreed that we should be winkled out from the comfortable work stations of our beloved Hudson and thrust kicking and screaming into the cramped and very uncomfortable Bisleys of 614 Squadron.'

A crash conversion-course on it quickly brought home to the newcomers that the designers of American types, like the Hudson, clearly had more concern for crew comfort than the British. He recalled that getting into the back end of the aircraft was 'the easy bit, but getting into the turret required gymnastic ability'. He told how 'one was required to curl himself up like an eel, and squirm up through an aperture of miserably small dimensions, with bits sticking out all around to snag on one's parachute harness and Mae West.'

Once in position the seat – similar to that of a bicycle – was folded down for the gunner to perch on for hour after hour.[3]

On the squadron's first day of operations, 6 June, three sorties were flown to escort convoys. One aircraft suffered the perennial Bisley failing – a burst tyre on take-off – but managed to get airborne and complete the mission. Another failed due to engine trouble, while a third stalled. It was not an auspicious start. The next day Bisleys flew on anti-submarine patrol and others as convoy escorts. The squadron diary duly recorded the results, 'No submarines reported and nothing unusual occurred.' The next day's operations were to have a similar result, 'No submarines sighted. Nothing of further importance to report.'

This was the lot of the maritime patrol aircraft; to scan endless stretches of ocean, often with no tangible results. The mere presence of patrolling aircraft was often enough to deter or disrupt submarine attacks which, of course, would never be seen. However, coming from the intense period of a high-tempo operation in the final weeks of the Tunisia campaign, the dull hours over the Meditteranean must have come as a jarring change to the now battle-hardened aircrews of 614.

But there were diversions. To gain insight into submariners' work, there were visits to Royal Navy and French ships and submarines for exercises, and practice bomb attacks were made on US destroyers *Plunkett* and *Giles*.

There was football against other squadrons. One game, against No. 813 Naval Air Squadron (NAS), who flew Swordfish biplanes, ended in a 2–2 draw, with the ground and aircrews 'showing great enthusiasm', the diary recorded.

One crew caused consternation when returning late from a sortie. No message had been received from the aircraft and a Swordfish of 813 NAS was duly sent out to search for the missing Bisley. It later returned safely but the crew faced a reprimand for failing to send an 'overdue' message.

The squadron again fell foul of proper procedures when the crew of F/O Aubrey-Rees' Bisley thought they had spotted a U-boat and launched an attack. The 'U-boat' turned out to merely be an oil slick. A warning was later given to all crews that oil slicks must not be attacked 'unless evidence of a submarine is seen'.

The sense of frustration was compounded on 10 July with news that the Allied invasion of Sicily had begun. The tone of the squadron diary is generally brisk and upbeat, and criticism of wider strategy is rare. However, it noted that, 'General opinion of aircrew is that it is a pity that the squadron's experience was not used in the operation, having regard to our work in Tunisia,' adding, 'Crews are obviously bored with this peaceful life after the action in the Tunisian campaign.'[4] There was recognition for service in that campaign on 26 July when both Flt Lts David Smyth and Derek Wallace were awarded the DFC.

However, an incident in the coming days would reveal the frustrations felt by many of the crews in a way that was hazardous and hilarious in equal measure.

Flying on 28 July, Rex Thompson, with his pilot Walter Matthews and navigator Eric Turner, were briefed to search for a U-boat that had been sighted near the Balearic Islands. Their course took them close to one of the outer islands, and there, resting upon the water in a quiet bay, they saw a German flying boat.

Rex recalled, 'Walter called base over his VHF radio (instead of me trying to slowly send a long Morse code message), giving our position and our intention to attack.'

In the meantime, the crew of the Blohm and Voss flying boat, who had been enjoying a swim, were desperately trying to get out of the water in a frantic effort to man their battle stations.

The message came back from HQ – 'permission refused', as the flying boat was inside neutral Spanish territorial waters. Reluctantly they abandoned their plan and returned to their U-boat search.

Four hours later, the patrol was at an end and the exasperated crew approached the North African coast. Having identified themselves as a friendly aircraft by firing the correct flare, they crossed the coastline.

Once safely inside the protective anti-aircraft umbrella of the shore guns, without warning or explanation, Walter Matthews lowered the nose of the aircraft into a steep dive. The Bisley gathered speed rapidly, and by the time it levelled out just a few feet above the ground, Rex Thompson remembered, 'we must have been close to top speed for the old Bisley'.

In the meantime, Rex had swung his turret to face forward, so that his guns didn't obscure the view behind the speeding Bisley. Hurtling along very low, Rex watched from his turret with 'fascination and more than a little excitement' as they zoomed over the Algerian countryside scattering goats, donkeys and people in all directions.

Directly ahead, about a couple of thousand yards away, was a half-built haystack on top of which were two farmers.

He remembered, 'Suddenly, one spotted our fast-approaching aircraft which was actually lower than they were. For a second, one of the farmers froze, then he banged the other one on the back and together they both dived over the far side of the stack, just as Walter raised the nose of the Blenheim.' He saw two pairs of heels disappearing out of sight as the hapless farmers escaped and the Bisley flashed overhead.

By this time it was dusk. It also happened to be the night of the week that the camp at Tarafaoui had a film show. The venue was a screen rigged up outside, with the audience sitting on the ground. To fill in the time until it was dark enough to project the film, 'Housie-Housie' (Bingo) would be played. As Walter's Bisley approached their base, still at just above ground level, he lined up the Bisley's nose with the back of the cinema screen.

As they burst out over the top of the screen, Walter pulled the nose of the aircraft high into the air and applied full power. Rex looked back at a scene of carnage beneath him. 'There were men sprawled on the ground, diving and running for cover, with some frozen to the spot while "Housie-Housie" cards fluttered around in the general confusion.'[5]

Walter, too, remembered the incident very clearly. After being denied the chance to attack the U-boat, he recalled, 'I took my steam out by beating up the camp on our return at dusk as they were assembled in the open-air theatre waiting for the movie.'

Unfortunately, among those waiting to see the film that night was the CO. By the time they landed and taxied in, there was a high-ranking reception committee in a staff car waiting to greet them and they were grounded until 11 August.

Walter recalled displeasure at his stunt went all the way to the top, 'I got red ink on the orders of Air-Vice Marshal Sir Hugh Lloyd [AOC NACAF].'[6]

By 14 August, word came that 614 was to move, but the destination was 'most secret'. On the same day, the squadron suffered its final casualty in North Africa when LAC Frank Edward Porter was killed when his lorry overturned.

It was also confirmed that Wg Cdr Bonner was to become the CO of the squadron after Wg Cdr Sutton was admitted to hospital with what was called a 'poisoned leg'.

On 19 August, their new home was disclosed as Borizzo in newly invaded Sicily. The main road convoy left for Tunis, while a small rear party remained at Tafaraoui maintaining aircraft. Some aircrews also stayed and continued with carrying out convoy escort duties. An advance party left on 21 August in two DC3 Dakota transports and had landed in Sicily by 18.00. No. 614 had arrived on the continent of Europe and Italy would be their home until the end of the war.[7]

At Borizzo, an 'excellent site' for the squadron's camp was found in an olive grove with a good water supply from local wells about two miles from aerodrome.

The locals were reported to be 'friendly and co-operative' and arrangements were made locally for laundry at 'reasonable prices'.

Furniture and timber was scavenged from bombed-out houses at nearby Marsala. As work on the campsite progressed, the airfield was improved with American engineers levelling and extending the runway. A bomb was also removed from a crater at the site, and the Americans loaned a jeep for communications. By 27 August a signal was sent to CO 614 Squadron that the sites were ready. With the campsite completed, half of the squadron at Borizzo were taken to the nearby beach 'for bathing'.

The final sortie from Tafaraoui took place on 3 September and the final Bisleys departed North Africa. The main party made the long journey by road to Tunis and then travelled by sea to Sicily. It was confirmed that the squadron would now be tasked with air-sea rescue.

Training on their new air-sea rescue role included lectures on wireless (W/T) procedure and Lindholme dinghies (specially-designed lifeboats dropped by parachute).[8]

On the day the uneventful convoy escort duties ceased in North Africa, some excitement returned to 614. On 3 September, the surrender of Italy was announced and Walter Matthews, with Rex Thomspon, were detailed to fly to Cagliari in Sardinia to receive the surrender of a fighter squadron there.

Arriving in Cagliari, the Italians were paraded in front of their Macchi 200 fighters which were all parked in a perfect row. In a short 'but touching' ceremony, the Italian Squadron Commander surrendered his credentials of authority to Walter.

The formalities complete, and the parade dismissed, all the pilots allocated to an aircraft were called together for a briefing for the return flight home to Sicily.

After checking that all weaponry was unloaded and all ammunition magazines were empty in the fighters, Rex Thompson asked the Flight Commander if he would fly close by to them during the return flight so that he could take some photographs. The sleek Italian fighters had their markings painted and replaced by roundels on each side of the fuselage.

After they took off, the Italian Flight Commander eased his aircraft close to the 614 Bisley on the return flight, and Rex took his pictures.[9]

As operations stepped up, with huge amounts of air traffic for the invasion of the Italian mainland, the squadron prepared to take on its new role. An air-sea rescue specialist, Sqn Ldr Collins, was attached to the squadron temporarily, and he organised the setting up of an operations room focused on their new task.

By 7 September, all of the squadron was at Borizzo, and two days later had made its first air-sea rescue sortie. A new chapter for 614 had begun. There would soon be high drama, in contrast to their previous role.

In an early-morning search on 13 September, F/O Johnston's Bisley spotted an empty dinghy. Later, F/O Matthews' crew spotted an enemy submarine which they pursued but it crash-dived and disappeared below the surface before they could attack it. Although not equipped with depth-charges, as they would have been in their former role, they could at least fire on the vessel with their machine-guns – but the submarine vanished before they could do this. Following this, F/O Flynn's crew spotted a 'burning hospital ship' while searching for the crew of a downed aircraft.

The last sortie of the day left Borizzo at 14.08, with Flt Lt Holloway as pilot. As with most of 614's missions at this time, their objective was to search a set area where it was thought an aircraft had ditched.

Ted Holloway had only joined the squadron in June, having been transferred from 18 Squadron. He had flown alongside 614 Squadron in one of the Bisleys which took part in the disastrous raid on 4 December which had seen a posthumous Victoria Cross awarded to the leader of the raid, and all aircraft shot down. He and his crew were lucky to survive that day, but it seems on this Monday, the 13th, his luck was to run out.

The sombre note in the diary read simply, 'Flt Lt Holloway failed to return from sorties. F/S BT Harvey (Nav) and F/S GL Taylor (WOP/AG) are the two other members of the crew.'[10]

Rex Thompson remembered how it could easily have been him rather than F/S Taylor taking the WOP/AG seat on that fateful mission.

He had flown with Ted Holloway before, and was full of respect for his skill as a pilot. Indeed, he credited him with saving his life.

He recalled, 'I had flown with Flt/Lt Holloway on a previous occasion. When not far south of Stromboli, one of our engines began to falter. With all that sea below, separating us from base, we headed homewards. To keep our old girl flying, he requested we jettison everything we could lay our hands on – that is, everything except our precious inflatable rubber dinghy. The fact that I am still here writing these words is a testimony to the skill of Flight Lieutenant Holloway as pilot of our sick aircraft, for he gently coaxed her all the way back to base, terminating our sortie with a perfect landing.'

He told how this piece of flying gave him enormous faith in him as a pilot. So, when, the following week, on the morning of 13 September when Holloway came to their tent and asked if he would take over from his sick WOP/AG, without hesitation Thompson accepted.

He joined Flt Lt Holloway and navigator F/S Harvey for the briefing in the operations caravan. They collected their gear and started walking out to the aircraft. But others had different plans for him that day, he recalled, 'We hadn't gone many paces when there was a bellow behind me, "Thompson, where the hell do you think you are going?" It was Sqn Ldr Watson, Rex's pilot.

'"On an op with Flt/Lt Holloway, Sir."

'"Like hell you are. Didn't you read DROs [daily routine orders]? We are on standby crew."'

So, as Rex remembered, 'That was the end of my flight with Holloway, who then had to backtrack to our tent lines and find another volunteer in my place.'

Being standby crew there was little else to do but hang

around the operations caravan all day in a state of readiness in case they got a call-out – so they were on hand when Flt/Lt Holloway failed to report in and failed to return at the time he would have been expected. He remembered the anguish that spread around the van.

'Another hour passed and still no sign and no sound over the radio. Things did not look good. It was too late in the evening to mount a search. An inflatable is hard enough to see in daylight, so a night search would be impossible.'

At first light in the morning several aircraft were dispatched, which searched 'every inch of water in our allocated search area, plus the intervening sea back to our base'.

He said there was 'a real urgency and determination amongst all of us, to do everything we knew how, to locate survivors'.

Searching continued and, on 16 September, the diary told how Flt Lt Holloway and crew were 'presumed missing'. The next day two aircraft made a final search of the area to which Holloway and his crew were sent. The diary sadly recorded that 'a dinghy was found containing no occupant'.

Rex recalled sadly, 'War is war and we expected losses, but somehow this seemed particularly upsetting since there was no knowing what had been their fate. The popular scenario being, that, since there was nothing heard from them over the base radio, they were probably jumped by enemy aircraft and shot down before a message could be transmitted.'[11]

Meanwhile, other sorties continued and techniques were changed with aircraft flying in pairs to search a larger area and make a more complete search. The promise of new aircraft was again dangled in front of 614 when the AOC visited the squadron on 19 September. He outlined the progress of the war and told them that air-sea rescue would soon be taken over by specialised Warwick aircraft (the air-sea rescue variant of the Wellington bomber) and that 614 Squadron would soon revert to anti-submarine patrols for which they would be re-equiped with the American-built Venturas in the following few weeks. But, like other prospects of newer and better aircraft that had appeared before, they never materialised and the squadron continued to soldier on with the ancient and unloved Bisley.[12]

The long searches across the sea were occasionally livened up in other ways. Rex Thompson recalled one sortie where they had been out for about three hours searching for a U-boat in an area south of Capri. He was asked to relay a coded message back to base. 'To do so required a fair amount of physical manipulating to reach the radio and the morse key from inside the turret, which in turn meant that one's attention was diverted from where it should be – searching the surface of the sea and searching the sky for aircraft.' While thus occupied, without warning, there was a frightful crash on the turret which sounded like a bullet passing through.

'My head snapped up and, to my horror, there were two fighters in formation closing fast.' He assumed they were under attack from the pair. 'I called the pilot and reported their presence and to standby for me to give evasive action instructions.'

As the shock gave way and he got the fighters firmly in his gunsights, he was able to see them more clearly. 'It slowly dawned on me that they were Bell Airacobras [American-built fighters] belonging to the Free French Airforce. Our Blenheims looked something similar to a Ju88,' he admitted, 'so I guessed that they were checking us out, that's why they kept station with us but just out of range of my guns. I reported to the skipper that they were now identified as friendly, and now relaxing a little I glanced around the turret to see what had made that frightening crash.'

The cause of the panic – of the adrenalin coursing through him as he thought they were about to be shot out of the sky – was both simple and a little ridiculous.

'All I could see was a slab of Spam, about the size of the

palm of my hand, glued to the perspex of the turret about three inches from my right ear. I found out later that the skipper, on opening the lunch prepared for him by the officers' mess, believed that the Spam had "gone off", so he peeled it from his sandwich and flipped it out of his little sliding window. The slipstream did the rest – smashing against my turret.' Bizarrely, at the very moment the unwanted slice of Spam had been propelled at 150mph to collide with his turret, the French fighters had appeared. Hence Rex's frantic reaction.[13]

By late September the Allies had advanced onto the Italian mainland. However, there were believed to be Germans still on the island of Sardinia but that they were retreating and preparing to evacuate to neighbouring Corsica. The situation was ambiguous and potentially dangerous.

Walter Matthews' crew again would find themselves involved in an unlikely incident. On 29 September, along with the Bisley of Flt Lt Robarts, Officer Commanding 'B' Flight, he was tasked with a search and, if possible, the dropping of dinghies to an American crew reported to have been shot down off the north-east coast of Sardinia. Bisleys 'R' and 'Z' duly took off for a round trip of about 700 miles, plus the time spent in the search area.

But while on transit to the search area, things began to go wrong. Walter Matthews recalled seeing something 'quite large' fall off Robarts' Bisley ('R') and drop into the sea. This, he learned via radio message, was the propeller which was followed by other parts of the Bisley's radial engine. The aircraft, Walter recalled, 'was not noted for its single engine performance'. Robarts headed his stricken aircraft towards Sardinia. His crew jettisoned everything that wasn't essential in an effort to lose weight and keep flying as they went along. 'Even if the Germans had not vacated that area,' Walter said, 'it was a more desirable option than to attempt the long return flight back to base. That would almost certainly terminate in ditching.'

Abandoning the search for the Americans, they escorted Robarts to the island. 'We headed for an airstrip that we knew to exist on the northern end of Sardinia, called Terranova,' Walter said. Meanwhile, Robart's aircraft was gradually losing altitude and they were gravely concerned they would not make the distance. They were greatly relieved to watch him safely make a difficult wheels-up landing on the bomb-cratered runway.

Following Robarts, Walter (flying 'Z') made a few orbits and some low passes of the airfield to try to find a spot to land among the bomb craters. Having landed, he pointed the nose of the aircraft directly at the airfield buildings and told his WOP/AG, Jim McTamney, to keep the twin machine-guns in his turret trained on them, lest the enemy appear.

In the meantime, he checked with the other crew and found them all to be fine. While they were discussing their next move, a figure appeared from the buildings holding aloft a white shirt tied to a broom handle. He had a purple sash across the front of his shirt, and he explained that the sash identified him as the Italian adjutant of the station. 'He was,' the adjutant explained, 'surrendering the aerodrome to them.'

The Germans had departed a couple of weeks before, so they felt able to relax somewhat. Robarts took Walter's Bisley ('Z') and headed back to Borizzo, having removed equipment from the stricken Bisley and returned to base at 16.50. Walter, Rex and Jim, with the help of plenty of willing Italian helpers, set about filling-in enough craters to make the runway suitable to fly out a heavily-laden aircraft when they came to pick them up. They also removed all the secret equipment from the damaged aircraft and destroyed some other items.

That night, as they waited for their rescue flight to come, a goat was killed and they gathered green salad leaves from the surrounding countryside. New Zealander Walter described them as similar to the 'Puha', which is like a Kiwi version of the dandelion. Sated by their rustic roasted goat and greens, they waited for their rescuers to return.[14]

There were further twists, however. The next day, Wg Cdr Bonner in Bisley 'X' and Flt Lt Smyth in 'F' flew to Terranova to pick up aircrew left there. One can imagine Walter's relief at the appearance of their rescuers. Unfortunately 'X' was unable to take off again, so Wg Cdr Bonner returned in 'F'. There were now two unserviceable 614 Bisleys at Terranova. The following day a further rescue bid was mounted with Flt Lt Smyth in 'D' and F/O Prestney in 'H' leaving for Sardinia with groundcrew to service the stricken 'X' and to bring back aircrew. Happily, they were successful and, on 3 October, the three aircraft 'D', 'H' and 'X', returned to Borizzo safely.[15]

A week or so later, on 11 October, the squadron had some news which while welcome, would leave many of them badly shaken.

All squadron personnel were instructed to gather at 10.00 as Wg Cdr Bonner wished to address them all. There was speculation that they were being posted to Blighty or maybe a new role was being planned for them. But the news he gave, Rex Thompson remembered, caused 'many tears and sad faces'.

The CO climbed on to a chair so that he could be seen by all, and opened a letter that he had just received, and wished to read it to the squadron.

The letter was from Flt Lt Ted Holloway – the man that everybody believed was dead. His Bisley had disappeared, seemingly with all the crew while on a search near Salerno the previous month. As already noted, Rex Thompson had narrowly missed being part of his crew that day.

'That created a stir among us to learn that he was alive,' Rex recalled years later, adding, 'The words in the letter touched me so deeply that I am confident that even after all the intervening years, I can still give an accurate account of his letter.'

Ted Holloway explained in the letter that he was writing it from a hospital bed in Africa. (His serious condition, together with the unpredictable mail delivery, would account for the great length of time in hearing from him.) Ted remembered completing his search and turning for home – his next conscious recollection was being underwater looking up to the light on the surface above – half-drowning and choking. He surfaced to find that he was utterly alone – no aircraft, no other crew members – nothing. He inflated his Mae West life jacket and rested a while, in a complete state of shock.

Not far away a stream of bubbles started to erupt on the surface, to be shortly followed by the partly-inflated aircraft rubber dinghy.

The CO broke off at this point to explain (as they all knew) that the only way that the dinghy could have surfaced was for the gunner to have opened the hatch on top of the fuselage and thrown it out. The jerk on the cord attaching it to the aircraft would activate the CO2 bottle to inflate the dinghy and take it clear – but the aircraft must have sunk very fast not to give the WOP/AG time to get out himself, or maybe he was injured and could not get out.

Rex recalled, 'Obviously, the dinghy must have gone down with the aircraft, but, as it inflated its buoyancy snapped the attachment cord allowing it to float to the surface.'

Ted wrote how sad he was at having lost his crew and that he would willingly have changed places with them if it had been humanly possible.

He wrote how he had then started swimming towards the safety of the half-inflated dinghy but, as he leaned forward to swim, water started to drown him, and it was only then that he discovered that he had a hole in his windpipe, through which the water entered when he tried to move.

When he finally made it to the dinghy and attempted to clamber on, all he succeeded in doing was to force the remaining air from the inflated part out through a large tear, so he had to give up that idea, and use the remaining flotation to cling on to.

Incredibly, having miraculously survived this ordeal, despite an injury which could have easily caused him to drown, there was still more desperate struggle if he was to survive.

He wrote of how during the night a 'large fish' came in very close to him as he floated in his dinghy. He had believed that there were no sharks in the Mediterranean, but when it came in several more times he succeeded in stabbing it on the snout with his pocket knife, so he said that it was close enough for him to identify it as a shark. Shark or not, it eventually left him alone to wait for the dawn.

Thankfully, later that day, he was picked up by a Polish destroyer, the *Piorun*, and taken to North Africa.

He explained that he had a complete memory block as to why the aircraft had hit the water. In the absence of a satisfactory explanation, he agonised continually as to whether it had been caused by any fault of his making. He did speculate, whatever the cause, the aircraft must have contacted the water with great force, for he believed that he had been catapulted out through the canopy which would account for, and was consistent with, the many injuries he had received. If that was true, the navigator, whose workstation is forward in the glazed nose cone, would not have stood a chance.

Years later, Rex confessed how he was also troubled by what he called 'the "IF" side to this event'. 'Had my skipper, Sqn Ldr Watson, not seen me walking out to the aircraft, perhaps I would be at the bottom of the Med, or perhaps if Flt Lt Holloway had not required to backtrack to get another WOP/AG in my place, which in turn delayed his departure time, then the aircraft would have been at a different place at a different time with the strong possibility that nothing would have happened. We shall never know.'[16]

Following the news of Holloway's survival, the squadron made efforts to see the injured airman and, on 12 October, Sqn Ldr Robarts flew to Setif in Algeria to try and find him. It was believed he was in 103rd Hospital, near Châteaudun. A day later Robarts returned from North Africa saying that Ted had now almost recovered and was going in a few days to a convalescent hospital in Algiers.

The squadron diary entry for 13 September, the day Flt Lt Holloway's aircraft failed to return, was amended, with the addition, 'Pilot now known saved.'[17]

On 17 October, the squadron came under Air Vice-Marshal Keith Park, AOC Malta. Famously the former boss of No. 11 Group RAF during the Battle of Britain, he paid them a visit on that day. Wg Cdr Bonner called the aircrew together and 'gave as much information as to the future of the squadron as possible'. They would continue in the air-sea rescue role for now, but for how much longer was not clear.

That winter was livened up by a search for meat and vegetables for their Christmas dinner, which they foraged from a local farm. One of those who took on the job of foraging was Scot, Willy Fraser. But one night, when they were about to go out hunting and gathering, a disquieting incident stopped them.

Willy Fraser recalled, 'A couple of us used to go round and buy, barter, or steal chickens, ducks, and geese for the squadron's Christmas dinner. I used to do the throttling, plucking and dressing of the birds for the officers' and sergeants' Mess.'

Sat in their tent in a Borizzo olive grove, Jock Grierson, Robert Glover (known as Dobbin) and Willy heard a crash out on the Trapani to Marsala road.

'We rushed out to find a horse and cart upside down and two men lying on the road. The crash had been caused by a stolen R/T Signals van that had been returning from Trapani. The signals van ended-up down an embankment and no driver was to be found. With torches we searched the crash site and found two Sicilian farmers in a very bad way. The biggest of the two was a man of about 15 stone, and the other around nine stone. I sent for medical help and a corporal arrived on the scene.

'Before he arrived I checked the big man lying in the road. There was no movement and I couldn't find a pulse in his arm or neck and I shouted to my friend, Jock Grierson, "The big yin here is dead", and a blanket was found to cover him up. The small farmer was creating hell with us and I assumed that his injuries were the reason for his gesticulations and noise.

'While we were attending to the farmers, their horse had been lying there groaning with frothy blood bubbles coming out from its throat. I took a close look with a torch and found that the end of the cart shaft had penetrated the windpipe and left it wide open. The only humane thing to do was to put a bullet in its head to put the poor thing out of its misery and agony. This I duly carried out with my Sten gun. It was through an interpreter that, I discovered, my act of shooting his horse caused the wee Sicilian farmer to go off his rocker. Little did he know that his poor horse was done for.'

The squadron ambulance arrived on the scene and Dobbin and Willy got hold of a stretcher.

'We rolled the big farmer onto the stretcher and covered him over with a blanket. It was quite a struggle to get him into the ambulance. Dobbin had one end of the stretcher and was actually inside the ambulance when, just at that moment, the big farmer sat up on the stretcher with the blanket over his head. Dobbin and I got the biggest bloody fright of our lives as the big sod fell on top of me. Jock Grierson shouted at me, "I thought you said he was dead". This caused some hullabaloo among the guards and other people who had arrived on the scene.

'Next morning, the MO sent for me and had a few words

to say. He did give me a real bollicking for declaring a man dead.'[18]

Christmas Day itself saw two convoy escort patrols, but the groundcrews had the whole day free. A soccer match was played between squadron teams during the morning on the aerodrome. Christmas dinner of turkey and plum pudding was served to airmen by the officers. A concert was held in a large marquee, with 'excellent entertainment provided by squadron personnel' and the day ended with parties in officers', sergeants' and airmen's Messes.

The year ended with three aircraft carrying out air-sea rescue searches for a missing Wellington, but with no success. As Borizzo was lashed by 40mph gales and rapidly deteriorating weather, there were new year's festivities after duty. A concert was given in the airmen's marquee, followed by community singing.

The new year didn't bring any relief from the dull ASR work. During January, the squadron had begun to take on more and more radar calibration flights. Monotonous, but necessary, they involved evaluation of navigational aids and electronic signals, to ensure they were safe and accurate. Often, calibration flights were combined with ASR missions.

The squadron had some interesting visitors. A representative from the Bristol aircraft company, one Mr Knapp, arrived at Borizzo. It is not clear what the purpose of his visit was. Wg Cdr Bonner departed to Malta for a brief visit on 15 January, and a week later the AOC AVM Scarlett-Stratfeild visited the squadron and addressed aircrew. The next day, 614 flew what would be its last sortie in the Bisley. By 22 January, the CO flew to Malta and back the same day and revealed that the squadron would be re-equipped and on six hours' notice to move. The next day saw all sections packing up equipment and handing it over to the Wing. Aircrew returned all their flying equipment. The squadron was confined to camp 'all ready for the movement'.

The squadron had another important visitor on 26 January, when Air Vice-Marshal Sir HP Lloyd came to Borizzo. Addressing the squadron, he told them of its valuable contribution since landing in North Africa, and the importance of air power in 1944. He wished them all the 'best of luck'. The next day, ground personnel (cooks, drivers, armourers, etc.) were posted away from the squadron and orders to move were awaited. It must have seemed that there would be little of the squadron left to move. Personnel were posted away, aircrew handing back their flying equipment.[19] What would their future be?

In the end a telegram from AHQ in Malta confirmed the squadron's fate. A series of meetings and visits had discussed the future role of 614. The message killed those discussions stone dead.

Marked PERSONAL, the message, on 30 January 1944, for W/CDR BONNER, FROM AOC (AVM Sir Keith Park) read:

'Coastal Air Force has signalled me that all postings by MAAF [Mediterranean Allied Air Forces] must stand. They have made it quite clear 614 is being disbanded.'

In a personal note, he added, 'That is final. Sorry you could not have gone home as a squadron.'[20]

Despite the sympathy from the AOC it meant that 614 had reached the end of the road.

Although it was a sad occasion, for many it meant one thing: they would be going home – or at least somewhere away from the dangers and discomfort of war-torn Italy.

Rex Thompson summed up his and of many others' feelings, 'To be heading back to England was just wonderful, and we were like a bunch of kids as we bundled ourselves into the luxury of our first-class cattle truck – smelly, draughty, with only the floor to sit on. We could put up with anything though – we were heading home.'

Jim Redfern recalled the moment when the lorries carrying 614 went through the gates of Borizzo Airfield for the last time. 'One of the most touching moments I ever witnessed in my service life was to see Wg Cdr Bonner standing on the embankment and facing the road – he saluted every driver as they passed. It was a scene I shall never forget.'[21]

Charles Bonner wrote to Air Cdre Richard Cadman in Cardiff – still 614's Honorary Air Commodore – expressing his dismay at the disbandment. 'The blow was very bitter for us all, especially for the Auxiliaries and one or two of us who have been in the squadron for a very long time.' The Air Commodore replired that he was 'very much surprised and disgusted' at the news and would be taking up the matter with 'the Air Marshal'. His protests were sadly in vain.[22]

Rex Thompson remembered standing at the ship's railing the evening he left Sicily, watching the full moon rise out of the Mediterranean. 'A peaceful scene broken only by the silhouette of partly-submerged ships that fell victim to bombing raids.'

That peaceful departure from a Mediterranean port contrasted with their arrival in November 1942 – 14 months earlier. Then, the Germans and Italians were still fighting strongly in North Africa, the troopship that 614's main party had arrived on dodging bombs and torpedoes. Some who lived then were dead by now. Now Sicily and North Africa were largely quiet. Such were the fortunes of war.

It seemed this was the end for the squadron. Some, like Bill Hooper, had been members recruited six years earlier in pre-war Cardiff. Now they would be cast to the four winds and posted to other places. After fighting their own 'Phoney War', they had served with distinction in raids over northern Europe, North Africa and Italy. Was this where 614 Squadron's story would end?

CHAPTER 6

Italy, early days (1944)

THERE WAS A time after the last members of the squadron left Borizzo that 614 Squadron did not exist. Caught in a bureaucratic limbo, it ceased to be at the stroke of a pen (or the clack of a typewriter) in January 1944. But, even as the clerk typed out 614's death warrant in Malta, there were stirrings elsewhere which would lead to its rebirth.

No. 462 Squadron of the Royal Australian Air Force was formed in Egypt in September 1942 from detachments of Nos 10 and 76 Squadrons of the RAF. The RAAF wanted to have more Australians in the squadron, but had had little success. By August 1943, only a quarter of its 600 personnel were Australian and the RAAF requested it revert to the RAF.[1]

The squadron was part of No. 205 Group RAF, flying long-range, heavy bombers like 462's Halifaxes, on missions against targets across North Africa and the eastern Mediterranean. Part of the North-West African Strategic Air Force, it began moving from North Africa to Italy at the beginning of 1944 as the war progressed, and became part of the Mediterranean Allied Strategic Air Force (MASAF). The groundcrews of 462 Squadron left their base at El Adem in Libya on 1 February, arriving at Celone on the Foggia plain on 1 March.

In North Africa, aircrews were to familiarise themselves with 'secret equipment... devices which had revolutionized the night actions of Bomber Command'. In November 1943, a number of crews had been sent back to the UK for training in the operation of H2S radar. This was in preparation for the squadron to assume a possible 'Pathfinder' role when it moved to Italy, and they brought a few of the sets back with them to start training the squadron.

The Halifaxes of No. 462 Squadron left El Adem appropriately, as things turned out, on St David's Day – 1 March 1944.[2]

As they flew, by another bureaucratic sleight of hand, No. 462 Squadron had already been renumbered as No. 614 Squadron. The Welsh squadron, disbanded in January, would fly again. Although the original members from the Auxiliary days in Cardiff were now gone, the reformed unit would build a new *esprit de corps* while still proudly wearing the crest of the Glamorgan squadron. They would still count Welshmen among its members, but would be joined by a veritable League of Nations of airmen. Rhondda-born pilot, Haydyn Jones, recalled with affection and admiration the array of people he met when he joined 614 in the spring of 1944, 'They were a real mixed bag: English, Welsh, Scots, Irish, Australian, South African [there would be Canadians too].' In particular, he fondly remembered his own Flight Commander, the 'illustrious DG Hayward... known affectionately by many people as "Dagwood". He was a tough Springbok with a magnificent physique and the courage of a lion.'[3]

It was still the only squadron of the RAF with a Welsh motto – 'Codaf i Geisio' – 'I Rise to Search or to Seek', and they would again rise to search – but this time in powerful, four-engined heavy bombers. The targets they rose to seek would be great industrial centres, oil refineries, railway marshalling yards and formations of enemy troops. For the first time in its existence, 614 would be able to pack a sizeable punch with the Halifax B MkII – a fighting aircraft on a different scale to anything 614 had flown before.

In the early years of the war the Halifax was the RAF's main heavy bomber. By 1943 it was being eclipsed by the Avro Lancaster which could carry heavier bombloads over a greater distance and was seen as more capable by Bomber Command chiefs. Nevertheless, the Halifax could still fly 1,100 miles carrying 13,000lbs at a top speed of 234mph.

It was an impressive sight, sitting tall on the ground, its nose thrust high like the prow of a ship. One of the factories producing it was the London Passenger Transport plant at Leavesden. A 1943 picture showed one of the double-decker buses normally made there parked alongside a Halifax, its top-deck barely level with the bomb-aimer's position.[4]

But the bomber was not without its faults. To overcome problems with drag and excessive weight, manufacturers Handley Page gradually made improvements.

Some of the aircraft flown by 614 were inherited from No. 462 Squadron's B Mk II Halifaxes, which were locally modified to B MkII Series 1 (Special) standard. The mid-upper turret was removed and the nose turret smoothed over with doped [stiffened] canvas to save weight and reduce drag.[5] Others were the further modified B MkII Series IA, which had all but essential equipment stripped away, and the nose was then refitted with streamlined perspex. Improved Rolls-Royce Merlin XXII engines delivered 1,400hp.[6]

The squadron would fly a mix of the older machines which came with 462 Squadron and newer ones ferried with new crews from the UK. Later versions of the B Mk II Series IA standard were fitted with a rectangular tailfin after the earlier 'D'-shaped fin was found to sometimes cause uncontrollable spins when the aircraft was heavily loaded.[7]

Despite more powerful engines, they still weren't able to lift the Halifax higher than 20,000 feet, making them vulnerable to night-fighters and flak. Throughout 1944, the squadron would pay dearly for these limitations as they attacked well-defended targets across the Balkans and Central Europe.[8]

On entering the Halifax, the crew, wearing bulky parachute harness and pack, Mae West lifejacket, oxygen mask, flying helmet and sheepskin-lined flying boots over flying suit, would clamber aboard the aircraft through a small hatch behind the underside of the left wing. The tail-gunner would turn right to his position and the remaining crew members would climb forward over the front wing spar. The Flight Engineer's position on the left would be the first crew station to see. With a fold-away seat for take-off and landing, his work station was directly behind the pilot with a bulkhead between them.

Moving forward, on the left was the pilot's bucket-like seat to accommodate a parachute. From here he could grasp the control column, shaped like a figure-of-eight on its side. The massive array of throttle levers was within grasp of his right hand. Further to his right there was a step down to the wireless operator's position directly beneath the pilot. The navigator would sit at his chart table just in front of him with the bomb-aimer further forward in the nose.[9]

Although cramped, those unlucky enough to be shot down in a Halifax had some consolation, particularly compared to Lancaster crews. The Halifax's larger fuselage, with fewer obstructions, was easier to escape from. Of those aircrews who were shot down, 29 per cent of Halifax crews survived, compared with eleven per cent of Lancaster crews.[10]

One of those fortunate escapees was someone who would loom large in 614's story. The then Wg Cdr Don Bennett bailed-out from his stricken Halifax following a raid on the German battleship *Tirpitz* in Norway. The future Pathfinder chief evaded capture and walked to neutral Sweden.[11]

Meanwhile, in Italy, although the number '614' had officially been bestowed on the new squadron, it wasn't clear the change was realised at Celone. Although glad about the arrivals from Libya, the squadron diary tells of confusion about their future.

'The reunion of the air and ground parties was cheering for all ranks,' it said, but, 'In spite of the fact that we already had the No. 614 Squadron official stamp in the orderly room, when the Air Officer Commanding 205 Group first visited us, it was not known in 205 Group what our number would be, nor was it clear that we were to be a target-marking squadron, in spite of two radar officers, plus large quantities of H2S and GEE [radar and direction-finding] equipment being on the spot and awaiting our arrival and complete with instructions from the Air Ministry. This was rather a setback and strong representations were made to Group and anyone else who appeared to be in the slightest way interested, to clear up the situation.'

The 'strong representations' cleared the picture, the diary told. 'On 3 March 1944 we became officially No. 614 Sqn and although nothing definite could be given by Group concerning our future policy, it appeared likely that we would become a target-marking squadron.' (Records show the change happened on 15 February.)[12]

The diary of No. 205 Group also records this, but noted 614 would take on its new role sometime in the future. 'Information was received from HQ MAAF [Mediterranean Allied Air Forces] on 3rd [March] that No. 462 Squadron (Halifax) had been renumbered No. 614 Squadron and would, at a later date, become a target-marker squadron and pass to the direct control of HQ No. 205 Group from the operational control of No. 240 Wing.'[13]

Nevertheless, some 462 Squadron crews had been sent to the UK in November 1943 to train as Pathfinders, and returned to instruct others. No. 614 would be formed around 462, but supplemented by additional crews, trained as Pathfinders and flying from the UK.

That month they began to arrive and there was a 'a seemingly never-ending stream of TM [target-marking] trained crews and H2S aircraft arriving daily from the UK.'[14]

The crews from Britain were trained under the legendary, now Gp Capt Don Bennett, the founder of the Pathfinder force – who had returned to the UK after bailing out over Norway.

Bomber Command chief, Air Marshal Sir Arthur Harris, opposed the formation of this specialised group. Eventually, he was persuaded to drop his opposition by the Directorate of Bomber Operations and a number of Pathfinder 'enthusiasts'.

Notable among those enthusiasts was Bennett himself. An experienced long-range pilot and navigator, he was appointed to command the Pathfinders on 5 July 1942, prior to its official establishment just over a month later.

He set high standards, and sent crews which could not meet these high standards to main force (Bomber) squadrons, replacing them with hand-picked volunteers. He encouraged elitism and got authorisation for Pathfinder aircrew to wear a special badge below their normal flying brevet and medal ribbons. No. 8 (Pathfinder Force) Group was officially formed as a Group in its own right on 8 January 1943 – in time to lead Bomber Command's assault on the Ruhr.[15] No. 614 would not serve as part of 8 Group, but were Pathfinders nonetheless.

Navigator Dudley Egles was one of those who trained as a Pathfinder in the UK before flying out to Italy to join 614. Sporting a fine handle-bar moustache for much of his time in the RAF, Egles is someone for whom the term 'irrepressible' was coined. The Bognor man joined up in April 1940, aged 19, having repeatedly tried to do so from the start of 1939. After training as a navigator, he was posted to Egypt where he and his crew received a Mention in Dispatches for evading capture after force-landing behind enemy lines and making it back to British-held territory with the help of the LRDG, the early incarnation of the SAS.

He was posted back to the UK because of injuries sustained walking across the desert in flimsy flying boots, but quickly tired of his new role as an instructor and yearned to fly operations again.

His training started at RAF Upwood, where theory was taught. Here they learned target-marking and route-marking methods. Some of them were given code names inspired by the home towns of instructors like Don Bennett. Whanganui was the name of the method with so-called 'sky markers' or coloured flares. Parramatta was the name for ground-marking with long-burning incendiary bombs of different colours. Newhaven marking used H2S, but backed up with visual identification.[16]

Later they would be sent to RAF Newmarket (the name also famed for the nearby racecourse) for more practical Pathfinder instruction. Dudley Egles was told that after his training he'd be posted to a Halifax squadron in Italy to bring the navigators up to Pathfinder Force (PFF) standard.

Also at Upwood was a fresh-faced young Australian, Tom Scotland. The Perth man had joined up in Australia in 1941. Arriving in the UK in December 1942, he trained to be a bomber pilot on the Whitley at Abingdon in Oxfordshire and later graduated to the Halifax at Marston Moor in Yorkshire. It was at Abingdon he first met his crew – the men who would fly, live and fight together until the war's end. On meeting them, Tom had difficulty understanding his Geordie crewmates, and they his Aussie twang, but they soon gelled together.

There was navigator, Derek 'Hoppy' Hopper, a quietly-spoken Newcastle man, who trained as a navigator in South Africa. The wireless operator was Wes 'Shorty' Weekes, who was a fellow Australian. The self-deprecating Queenslander was modest about his radio skills but Tom rated him very highly. The bomb-aimer/observer was Edgar 'Ed' Riley. A Geordie, his experience as a draughtsman stood him in good stead with map-reading and navigation. The tail gunner was Bill Smith, from the Midlands. Later,

Flight Engineer Bob Lewis, from Glasgow, joined them at Marston Moor.

The Yorkshire airfield saw their friendship sealed in the most harrowing of ways. On Boxing Day 1943, they were practising evasion techniques with a Hurricane. Pulling out of a steep dive, the bomber began to shake violently. Looking out over the port side, Scotland saw one of the engines had torn loose. The wing was beginning to fall apart and flames licked the shattered engine. He ordered the crew to bail out, and Tom could have escaped too, but stayed in the burning Halifax. Steering it away from houses and factories, he crash-landed in a field, just avoiding a village, saving not only the lives of his crew, but countless people living beneath the stricken bomber's path.[17]

The skills, character and training of all the crew would make the difference between life and death as they went through training and on combat operations together. In the spring of 1944, they were selected to be trained as elite Pathfinder crew at Upwood and Newmarket. This band of brothers would then fly to Italy to serve with 614 Squadron carrying out long, dangerous missions over enemy territory where they would depend on each other completely if they were to survive.

At Newmarket, Tom Scotland fondly recalled their training when, on race days, the war seemed to stop for him and his crew. They would circle the course in their Halifax, letting them keep an eye on the racing. Watching with binoculars they'd shout encouragement to the horses below.[18]

Another newly-trained Pathfinder bound for 614, Haydyn Jones, recalled how, while at Newmarket, crews would learn the intricacies of H2S while flying over the Midlands.

'As residents of British cities heard aircraft overhead, they'd have been unaware their homes were being scanned by airborne radar.' Jones remembered that the target for many of their practice bombing runs – quite often above complete cloud cover – was Birmingham New Street railway station.[19]

Navigator Arthur Spencer joined the Pathfinders in the UK where he had completed a first tour of 45 operations with 97 Squadron. Although not officially a member of 614 Squadron, he went to North Africa to help train 462 Squadron and went with them to Italy as they transformed into 614 Squadron before becoming 205 Group's navigation officer.

He recalled that, at the operational training unit and at the heavy conversion unit, the emphasis was on the pilots, since they were learning to fly a much larger aircraft. When training to be Pathfinders, the importance of navigators and bomb-aimers was stressed. 'Pilots,' said Don Bennett, 'were merely chauffeurs to get the really important people – the navigators and bomb-aimers – there to put the target-indicators (TIs), flares and bombs down in the right place.' Arthur recalled that this was 'all very encouraging for the morale of those of us who were in these two trades'.[20]

Tom Scotland recalled that fellow-Aussie Gp Capt Bennett was tall, with piercing eyes and a commanding voice. The Pathfinder chief had good advice when he welcomed Tom's crew and five other crews to Upwood and to the PFF.

He urged them to, 'Get to know the heavens. Get to know the stars!' Tom said, 'I was ever thankful for the kinship I gained with the stars, basic to our navigating across continents.'[21]

As if to mirror 614's journey from Portreath to Algeria 17 months earlier, many Pathfinders of the newly-reformed 614 had to fly from the UK via the Bay of Biscay and North Africa. Another UK trainee destined for Italy was Hugh Fleckney.

His journey, littered with wrong directions, mishaps and narrow escapes, reflected the confused picture surrounding 614 Squadron's rebirth. Departing in early March, they'd been told, wrongly, that 614 was at El Adem. They left RAF Hurn, near Bournemouth, en route for Rabat in Morocco –

their first leg. It contained, Hugh recalled, 'the first of many frights I had during the flight over the Bay of Biscay'.

First, they took evasive action from a 'night-fighter'. 'This turned out to be the planet Venus. Then, when we should have seen the sea, we fetched up over what appeared to be a large fully lit-up town.' The 'town' was eventually identified as a Spanish tuna-fishing fleet.

There followed an odyssey across the Mediterranean as they tried to find somewhere to land and refuel and someone who knew where 614 Squadron was.

Arriving at Rabat in Morocco to find the aerodrome cloaked in fog, they diverted to a very crowded Gibraltar with aircraft parked on each side of the runway, resulting in a 'hairy' landing.

Next was Maison Blanche in Algeria, as 'someone told us that 614 had moved there'. This also turned was incorrect, and so they were told to try Catania in Sicily. The confusion continued, 'Unfortunately, it transpired that there was no 614 at Catania either but, at last, we were given firm assurances that the unit was now established at Celone in the Foggia plain.'[22]

Dudley Egles and crew had a similarly eventful journey when his flight hit trouble leaving Sicily. 'On the way from Catania to Celone we flew near Vesuvius, which was erupting. We discovered, to our horror, that our windscreens had become almost opaque – the volcanic grit had scratched the perspex and it was with great difficulty and skill that the pilot was able to land at our new base.'

Egles and other PFF crews from the UK soon began training the rest of the squadron. 'We soon settled in and fellow navigator Len [Palmer, with whom he had trained as a Pathfinder] and I commenced instructing crews on H2S and Pathfinder route and target-marking.'[23]

Even though 462 Squadron, which preceded 614, was transferred to the RAF because it had insufficient Australians, the newly-formed 614 still had many from Down Under. Tom Scotland met other Aussies, but some friendships would be short-lived.

Australian pilot Tony Wheeler showed Tom around after offering him space in his tent. The sports-mad Aussies made sure they had spaces to play when they weren't flying Tom recalled, 'Tony and I made a quoits court between the olive trees and arranged dusty, energetic team games with our crews.'

He discovered another Australian pilot, Jerry Atwell, and often made himself comfortable on the floor of his tent with his crew spread-eagled around chatting. Tom remembered, 'Jerry delighted us with tales of Australia and his love for the girl in the steel town of Port Kembla in New South Wales.'[24]

As well as the friendships between Commonwealth countries, the Americans were also neighbours, Dudley Egles noted, 'The Mediterranean Air Force was a combined British-American unit and Celone, like the other bases on the Foggia plain, was shared between them.' They found that Celone would be a spartan home, both for men and aircraft.

'We did not have a "hard" runway, and the metal strips (or pierced steel planking) laid down often buckled on the very soft terrain when heavy aircraft such as our Halifaxes and the USAAF Liberators and Flying Fortresses landed on them.' Accommodation was in tents. 'All very basic,' he remembered.

The Americans had better facilities. 'There were two American squadrons on the base. They, of course, had wooden floors to their tents, duckboard footpaths, and hot showers.'

These comforts were greatly valued by the lucky Americans because the weather was not at all 'sunny Italy'. 'It was wet, cold and very muddy.' Unlike the Yanks, the British had to fend for themselves. Most of 614's tents were fitted with an ingenious, and Egles thought, 'highly illegal' petrol drip-feed heating system made by groundcrew. 'It was crude,' he said, but 'did help to make life a little more bearable'.

The officers' Mess was a marquee with a 'magnificent bar' built by the unit carpenters and decorated by odd pieces of salvaged German aircraft, complete with swastikas.[25]

Hugh Fleckney, too, remembered the makeshift nature of 614's accommodation when he arrived. When he and his crew got to Celone, there were no living quarters available. For their first few nights in Italy they slept in their aircraft, dressed in flying kit because of the 'perishing cold'. 'In due course,' he remembered, 'we were allotted tents.' The tents, however, had no beds. 'We scrounged around for timber and string with which to make our own.' Food was basic but plentiful. All was tinned, including the 'Olio' margarine, which was designed to be used in warm climates but was rock solid in the cold of the Foggia plain.[26]

Meanwhile, on the ground the Allies were frustrated by the strength of the German resistance. By 1943's end, following their invasion of Italy, they were bogged down on the Gustav Line, a defensive barrier across Italy, south of Rome. Field Marshal Albert Kesselring – the German commander in Italy – took full advantage of the Apennine Mountains and narrow valleys of central Italy, all suited for defence. The landings at Anzio, south of the capital, in January were an attempt to bypass the Gustav Line and take Rome. However, they failed to produce a breakthrough.

In the wake of the Allies' struggle, on 19 March the MAAF launched 'Operation Strangle', an attempt to help the armies on the peninsula by interrupting and destroying the enemy's lines of communication. Marshalling yards in northern Italy were targeted by strategic bombers of the USAAF 15th Air Force and No. 205 Group which made up the Mediterranean Allied Strategic Air Force (MASAF) under the American Major General Nathan Twining.[27] No. 205 Group was made up of five Wings, each of two squadrons – four of RAF and one of SAAF. They flew Wellingtons, Liberators and Halifaxes, and 614 Squadron was within 240 Wing alongside 178 Squadron, who flew Liberators. They bombed at night while the 15th Air Force bombed by day. Some of 614's first raids would be in support of 'Operation Strangle'.

Over nine days, the squadron flew five operations which would be an intense and painful introduction. Before sufficient trained Pathfinder crews were available, 614's Halifaxes would operate as bombers rather than as target-markers. The results of those raids were mixed, and the cost to the squadron was high.

The first raid, on 11 March, while not a spectacular success, at least it went off without a major hitch. Seven Halifaxes attacked railway marshalling yards in Genoa, and although the target was obscured by cloud, reflection of fires or illumination was seen in cloud over the target. All returned safely to base.[28]

As well as raids on transport, 614's other early missions had a different objective. Four days later, the squadron would fly to a more distant target – a mission which would be one of its most costly. They would attack the far-off Bulgarian capital, Sofia.

The raid on Bulgaria had very different aims to those against transport targets. As early as 19 October 1943, Churchill suggested that the Bulgarians were a 'peccant [guilty] people to whom a sharp lesson should be administered'. Bombing was designed to 'undo the chord that bound Bulgaria to her German patron'. It was believed that the shock of these attacks would force Bulgaria from its alliance with Germany.[29]

The raids, which 205 Group began before 614 Squadron arrived, seemed to be working, and the Group diary recorded their effectiveness. 'The reported disordered and wholesale evacuation of 300,000 people undoubtedly led to considerable chaos in Sofia, in particular, and Bulgaria in general, and has helped to weaken the already precarious position of the Bulgarian government.'[30]

On 15 March, nine Halifaxes from 614 Squadron,

loaded with 500lbs and 1,000lbs bombs, flew the 400 miles over German-occupied Balkans to the Bulgarian capital with railway marshalling yards as the target. The raid was beset with difficulty. There was very bad weather with low temperatures causing icing.

Four 614 aircraft were known to have bombed the yards and several sticks of bombs were observed in the target area and city centre where a large, yellow fire was left burning. Severe icing caused several bombs to hang up, with one aircraft having all its bombs refuse to drop. Halifax BB385, flown by F/S Norrie Dear, was loaded with 7 x 1,000lbs bombs which it dropped in one stick. Bombs were observed to burst just north of the marshalling yards, with two sticks observed in the target areas. They got back safely to base, but others would not.[31]

Crossing the Adriatic, returning crews ran into dense cloud, with visibility down to zero feet with icing conditions. Several aircraft from the Group were unable to find their bases, and twelve of the force of 67 aircraft failed to return. Four of them were from 614 Squadron.

Halifax BB446 was thought to have been shot down over the target with no survivors. They were Pilot F/S Lyle Anzac Lawson, RAAF; Navigator Sgt James Martin Withers; WOP Sgt Frederick Henry Worthington; F/S David John Lewis; T/Gnr Sgt David Roland Williams and Bdr F/S Vivian Dickson, RAAF.

Halifax JN890 seemed to be returning and identified itself at 03.46, giving its estimated time of arrival as 04.10. Nothing more was heard from them following this but wreckage, later found on the shore north of Messina in Sicily, was found to be part of a Halifax. The words 'O/C 462 Squadron' were painted on the main wheel which would indicate that this was a 614 Squadron aircraft. There was no other trace of the aircraft or crew. Those lost were Pilot F/S Donald Herbert Jackson; Navigator Philip Walter Templer; WOP Sgt Walter

Lee; F/Eng Sgt William Frederick Hearn; T/Gnr F/S Robert Thomas Lightbody, RAAF and Bdr F/S GF Pridham.

Halifax BB382, flown by F/S Murray, almost made it back to Celone but crashed at San Nicandro around 35 miles from Celone. Three survived, but the other three were killed. They were Navigator F/O Henry Calvin Chambers Read, RCAF; Bdr F/S Donald Leonard Smith, RCAF and WOP Sgt Samuel John Boyce.

Another Halifax, BB415, didn't make it home, also falling victim to the foul weather, but through luck and perseverance the crew somehow all survived. Halifax BB415, flown by F/S RJC Dingwell, bombed the target and was returning to Celone when, around 100 miles from base, they hit bad weather and very low cloud. As they neared the Italian coast, they tried descending but were still in cloud. Contacting base they received QDMs (bearing to the station via radio direction finding) and confirmed they were near Celone. With enough fuel to last until daybreak, Dingwell decided to climb to 6,000 feet and circle until then. But when daylight came, the position was the same.[32]

Just days later another 614 crew, returning from a raid on a different Bulgarian target, also couldn't find Celone. Having attacked Plovdiv on 18–19 March, Sgt Hugh Fleckney and his crew were able to use another aid to locate home. Called 'Darky', the lost aircraft would call up on a special frequency and searchlights would be illuminated to point to the nearest airfield. Luckily, the weather was clear that night and the lights pointed them to nearby Taranto.[33] But Dingwell's crew, circling in thick cloud which the lights could not penetrate, was running out of options.

The wireless operator obtained two more QDMs and, using these, the navigator gave a course. Frustratingly, they passed over Celone but because of thick cloud they couldn't descend and flew to a pre-determined position which the navigator plotted. Dingwell recalled, 'On arriving there I circled at 7,500 feet and gave the order to abandon the aircraft.' All the crew bailed out and landed near the village of Melfi, 60 miles south of their base, which was deep in snow, where they remained until the weather cleared.[34]

It was a devastating loss at such an early stage in the reformed squadron's life but, as they reeled from this grievous loss, other crews were making their way to Celone from Britain.

Among them was Tom Scotland and his crew. They almost didn't make it and found out as they arrived that the squadron had believed them to be dead – and they were lucky not to be.

Having travelled hundreds of miles they neared their destination. On the penultimate leg of their journey they were flying over the island of Pantelleria approaching Sicily. Scotland gazed down at the island and suddenly heard what he described as a 'Whoooomph!' Something had exploded in the aircraft. Anxious shouts rang over the intercom. Bomb-Aimer Ed Riley saw the remains of the starboard tyre. Somehow it had burst with huge violence. Would they be able to land? Had anything else been damaged? They'd soon find out. On approach to Catania, Scotland remembered, 'The port side wheel touched the surface smoothly and began to take the weight of the Halifax. My work now was to hold up the starboard wing and to not allow the shattered wheel to grab the ground and dig in.'

When the wheel touched down 'something like a giant hand gripped the plane and started to twist it', Scotland recalled. The huge landing gear gave way beneath them, forcing the Halifax to skid sideways. Fuel lines ruptured and flames flashed along the wings.

Out of the corner of his eye, he saw fire tenders, foam, and men moving with hoses. The flames disappeared. 'Only an ominous spiral of smoke rose from the ocean of white,' he recalled. 'Quietness reigned. We had arrived.'

After recovering in Catania, they were flown to mainland

Italy before finally arriving at Celone by truck. A grim welcome awaited Tom and his crew. Accommodation was makeshift – mostly tents. Although he made them feel welcome, the 614 adjutant wasn't expecting them. He'd thought Tom's crew had been killed in a crash in the mountains of Sicily – such was the confused message that had reached them. Tom recalled, 'He said there were plenty of tents because they'd just lost four crews on an attack on Bulgaria.'[35]

Pairs of Halifaxes made the following three raids. The two on 16 and 18 March were against marshalling yards in Sofia and Bulgaria's second city, Plovdiv. On both occasions there were problems – either with guns or engines. The third was against port installations at Monfalcone in northern Italy when all aircraft returned to base.[36]

The Plovdiv mission was Hugh Fleckney's first for 614. He wasn't hugely impressed. 'The aircraft was pretty elderly,' he recalled, 'and we lost an engine just as we reached the target. Then, on the way home, we found that one of the bombs had failed to release, as the bomb catch had frozen. We got rid of it by pushing the heater pipe into the bomb-bay and melting the ice.'

A further raid by 205 Group highlighted the need for the Group to improve its target-marking capabilities. Perhaps this lack of expertise prompted the decision to confirm what 614's role would be.

Following the squadron's ill-starred beginning with the raids of mid-March, and the lack of clear direction from 205 Group on what the squadron's job would be, things had reached an impasse.

There were visits from staff officers, conferences at Group where further discussions were had following the AOC's visit.

Whether the indifferent performance of the first four raids was a factor, or the lack of clarity on 614's mission, or because of the lack of pathfinding expertise in the Group – a decision had been reached.

The diary summed up the situation, 'During the time we had been operating, no policy had as yet been arranged, but as a result of the AOC's visit and visits by RAF staff officers, and conferences at Group attended by the Commanding Officer, Wg Cdr Bill Russell, it was finally decided to stand down the squadron and concentrate on training.'

At least there would now be clarity on what job the squadron was to do, what was expected of it, and how it was to meet those expectations.

A key part of the policy was 'The squadron will be a target-marking Squadron for 205 Group,' and it fleshed out how this was to be achieved. 'It will consist of two operational flights of eight H2S aircraft each, with a training flight consisting of two dual non-H2S aircraft and four H2S aircraft. The squadron will be brought up to establishment in equipment and ground personnel. Non-essential bomber aircraft will be removed and the training flight will supply replacement crews.'

Like their counterparts of 8 Group in the UK, they would be an elite unit. 'These will be above average 205 Group crews with experience of night operations, and they will be converted on to Halifaxes and trained in target-marking technique on the squadron.'

There were some crews who'd already reached that standard (like Tom Scotland, Dudley Egles, Hugh Fleckney and Haydyn Jones) and some who had not. They would be trained up.

'In the meantime,' the diary added, 'we have 18 crews who have completed the PFF course at Newmarket (the majority of whom have little or no operations experience). Two raw Halifax crews with no experience whatever, four ex-Wellington crews converting to Halifax (with little or no operations experience and no H2S, Gee or Mk XIV [Bomb-Sights] night experience) and eight crews with operations experience on the squadron but no H2S, Gee or Mk XIV Bomb-Sights.'

It's almost possible to hear the sharp intake of breath that Wg Cdr Russell no doubt took. The diary entry ended, 'From this motley crew we are expected to turn out a first-class target-marking force.'[37]

A valuable insight into those early days when the squadron was learning its Pathfinder craft and paying a high price for it, is provided by experienced navigator Arthur Spencer.

From his arrival with 462 Squadron in March through to the summer of 1944, he helped train the squadron and joined crews on actual raids. 'I spent most of my time with the squadron's navigation officer and his team, talking to them about the techniques we had used in Pathfinder Force and the equipment we used.' Some of that equipment, he noted, they didn't always have in the Mediterranean, such as the Gee navigational system.

He recognised the difficulties the squadron was having in adapting to its new rolem, but was clear that this was not down to lack of skill or talent in 614 Squadron. 'Their first [raid] was something of a fiasco, not through any fault of theirs. This was a pity, because it took them some time to win back their credibility with the supporting bomber squadron.'[38]

The squadron diary explained how they could win back that credibility. 'Crews for training have therefore been divided into two groups, the "H2S trained" and "conversion to H2S [radar]", "Gee [navigation homing]" and "Mk XIV Bomb Sight".'

The H2S-trained crews were given a navigation test to ensure that they were up to standard. Once this was passed, they were given as much blind and practice bombing with the Mk XIV Bomb-Sight as possible. The remaining crews had a tough syllabus to complete, commencing with films on H2S, Gee and lectures. They then had to complete eleven exercises, starting with a demonstration by an instructor and going through various cross-country flights, with photo bombing

This shot of three 614 Sqn Hawker Hectors was taken over over Cathays Park in Cardiff on Empire Air Day, Saturday, 20 May 1939.

614 Sqn archive

WESTERN MAIL & SOUTH WALES NEWS, SATURDAY, MAY 20, 1939.

EMPIRE AIR DAY TO-DAY. [illegible]

This lineup of 614 Sqn aircraft was taken for the airshow at Pengam Moors for Empire Air Day on Saturday, 20 May 1939, and appeared in the *Western Mail* that day.

Western Mail

Pilots of 614 Sqn and Hawker Hector in front of the officers' Mess at Pengam Moors. In later years this became the clubhouse for St Albans Rugby Club before sadly burning down in the 1990s. Centre is the CO, Sqn Ldr Richard Cadman, and the other pilot not in a flying suit is the then Pilot Officer Norman Merrett.

614 Sqn archive

Westland Lysanders of 225 Sqn which was formed in October 1939 from the aircraft and personel of 614 Sqn's 'B' Flight.

Air Historical Branch, Crown Copyright

Bill Hooper and his wife Edith on their wedding day. She was from Grangemouth and they met while Bill was stationed their with 614 Sqn. Many from the squadron met their wives while stationed in Scotland.

Ian Hooper

Flt Lt Norman Merrett, who died in a training flight on 10 August 1940; he was one of the early joiners of the squadron in 1938.

Author's collection

Dai Park when he was with 614 Sqn, *c.*1939.

Diane Crawley

From left: Denis Ager, Ian MacLaren and Paddy Stapleton with Blenheim IV V5752 LJ E 'Edward' at West Raynham, prior to the first 1,000 Bomber raid on 30 May 1942.

Dermot Stapleton

Flt Lt John Scott who was awarded the DFC for his actions during the Dieppe raid in August 1942. Despite being horribly wounded, he flew his stricken bomber home in an effort to save the life of his Observer, his WOP/AG already having been killed when they were hit by flak fired from a Royal Navy ship.

Author's collection

F/S Paddy Stapleton.

Dermot Stapleton

The 614 Sqn football team in 1943. Centre of middle row is Wg Cdr Charles Bonner. By then based at Tafaraoui in Algeria, they played many games against neighbouring units.

614 Sqn archive

From left: Flt Lts Peter Le Cheminant, Bill Stewart (radio officer), the CO Wg Cdr Harry Sutton, Flt Lts Ian Maclaren, Derek Wallace, Peter Robarts and Sqn Ldr Charles Bonner before an 'op' at Canrobert in December 1943.

Charles Bonner and Peter Le Cheminant

Sgts Fred Bunting (left) and Ivor Self being marched from their Blenheim after crash-landing in a field near Aldeia Nova in Portugal en route from Cornwall to North Africa.

Fred Bunting

Armourers of 614 Sqn load a 250lb general purpose bomb into Bristol Bisley, BA783, at Canrobert on 20 January 1943.

Air Historical Branch, Crown Copyright

Second from left is LAC Henry George Hawken who was killed alongside LAC Leo Woodbine when the truck they were in was hit by a Bisley as it took off from Oulmene on 24 February 1943.

Judy and Felecia Cleasby

The 614 Sqn menus for Christmas and New Year for 1943–44 while they were stationed at Borizzo in Sicily. The New Year menu for 1944 featured places the squadron had been based.

614 Sqn archive

Picture from W/O Hugh Fleckney of the 614 Sqn Christmas dinner at Amendola. All were given a bottle of beer, funded by a donation of £25 from the County of Glamorgan Territorial Association.

Mike Fleckney

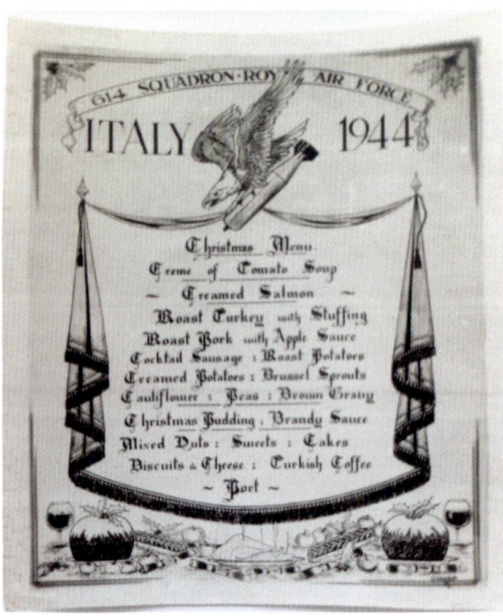

The menu for 614 Sqn's Christmas dinner in 1944.

Helen Mussell

The shattered nose of F/O Bruce Prange's Halifax, after the raid on Ploesti on 9 August 1944, was just part of the damage it suffered. All crew, excepting Prange and bomb-aimer W/O Al Fox, abandoned the aircraft after it caught fire following a night-fighter attack.

Author's collection

F/S Dennis Stanley walks through a lake of floodwater in front of one of the 614 Sqn's Halifax BMkII Series 1A bombers, JN976, after heavy rainfall at Celone, Italy, in March 1944.

Air Historical Branch, Crown Copyright

Accommodation at Amendola was very basic and the men had to improvise their own shelters, like this tent which has been improved with the addition of a front made from scrap wood. Picture from TNA AIR 25/829-819

The National Archive, Crown Copyright

ROYAL AIR FORCE
IDENTIFICATION CARD
УДОСТОВЕРЕНИЕ ЛИЧНОСТИ
DOWOD OSOBISTY
LEGITIMACE
LEGITIMACIJA
SZEMÉLYAZONOSSÁGI IGAZOLVÁNY
LEGITIMATION
Leslie J. Sims.
SIGNATURE

I AM A BRITISH AIRMAN
PLEASE TAKE ME TO YOUR COMMANDING OFFICER AND NOTIFY NEAREST BRITISH OR AMERICAN MILITARY MISSION IN BELGRADE, BUCHAREST, POLTAVA OR OTHER NEARBY PLACE. ALSO, PLEASE ARRANGE FOR TRANSPORTATION.
THANK YOU

SHOW THIS TO RUSSIANS
Я БРИТАНСКИЙ ЛЕТЧИК
ПОЖАЛУЙСТА ПРЕДСТАВТЕ МЕНЯ ВАШЕМУ КОМАНДИРУ
И УВЕДОМИТЕ БЛИЖАЙШУЮ БРИТАНСКУЮ ИЛИ
АМЕРИКАНСКУЮ ВОЕННУЮ МИССИЮ В БЕЛГРАДЕ,
БУХАРЕСТЕ, ПОЛТАВЕ ИЛИ ВДРУГОМ БЛИЖАЙШЕМ
МЕСТЕ. ТАКЖЕ РАСПОРЯДИТЕСЬ О ПЕРЕДВИЖЕНИИ.
БОЛЬШОЕ СПАСИБО!

Multi-lingual ID paper given to Les Sims of 614 Sqn. These were issued to aircrew to be used in the event of bailing out over occupied territory, enabling them to show resistance fighters or advancing Russian troops that they are friendly and should be given protection.

614 Sqn archive

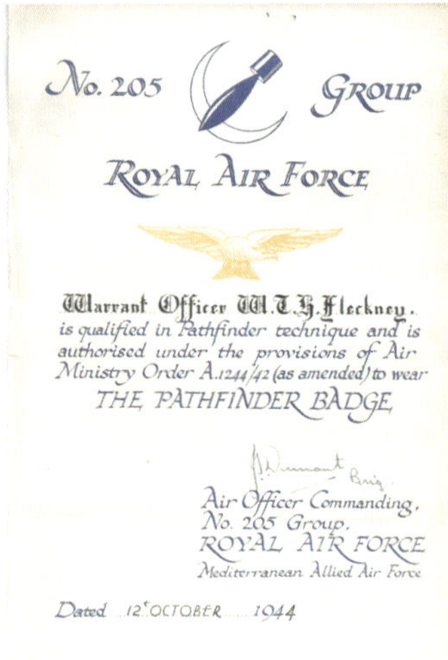
No. 205 Group
Royal Air Force

Warrant Officer W.T.H. Fleckney.
is qualified in Pathfinder technique and is authorised under the provisions of Air Ministry Order A.1244/42 (as amended) to wear
THE PATHFINDER BADGE

Air Officer Commanding,
No. 205 Group,
ROYAL AIR FORCE
Mediterranean Allied Air Force

Dated 12th OCTOBER 1944

Pathfinder Certificate awarded to W/O Hugh Fleckney.

Mike Fleckney

Shot taken by Wg Cdr Lawrence Moncrieff of a formation of Liberators.

Helen Mussell

Senior officers, including Brigadier Durrant, AOC 205 Grp, arrive for a presentation parade of Pathfinder badges and certificates to 614 Sqn aircrew at Amendola, November 1944. Note the rugby posts on the playing fields where the parade was held.

Mike Fleckney

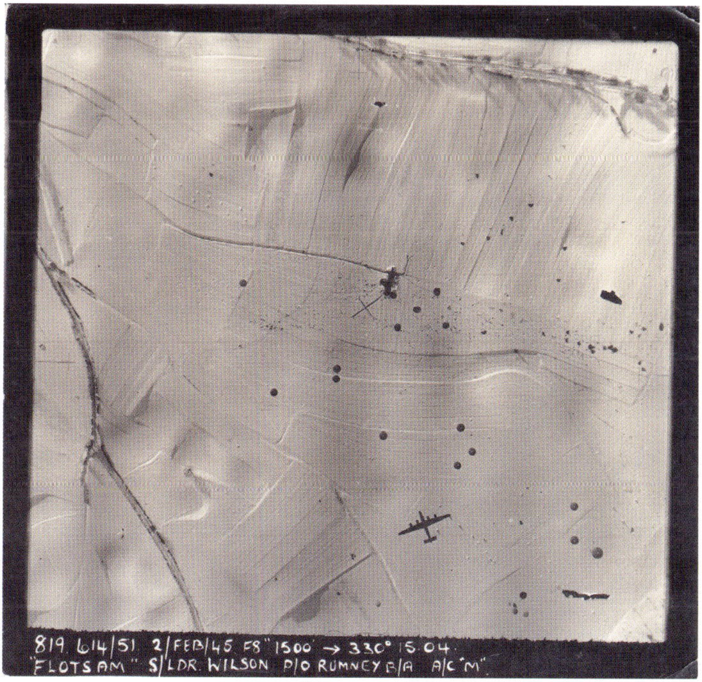

Shot of a 614 Sqn Liberator on supply drop, codenamed 'Flotsam', to Yugoslav partisans on 2 February 1945, flown by Sqn Ldr Wilson.

Helen Mussell

Wg Cdr Lawrence Moncrieffe pictured prior to the award of his DFC for his role in directing Pathfinding operations in 1945.

Helen Mussell

F/S MR 'Ron' Patriarca, who was shot down over Hungary, became a POW and was forced to march in dreadful conditions in 1945.

Alison Patriarca

Flt Lt Tom Scotland wearing both his Pathfinder badge and his DFC.

Peter Scotland

Navigator Dudley Egles trained to be a Pathfinder in England before joining 614 Sqn in Italy. He was shot down over Romania and became a POW in Bucharest.

614 Sqn archive

W/O Hugh Fleckney amid tent accommodation at Amendola.

Mike Fleckney

Picture of a 614 Sqn marking the target, Latisana Bridge, in northern Italy on 4 January 1945.

Helen Mussell

A Harvard trainer is fuelled up at RAF Llandow, *c.*1948.

Robin Irving-Lewis

Aircrew of 614 Sqn, including the CO, Sqn Ldr Bill Irving (with dog) and Spitfire MkXVI behind at RAF Llandow, *c.*1948.

Peter Irving-Lewis

Groundcrew refuelling a 614 Sqn Spitfire MkXVI at RAF Llandow, *c.*1948.

Peter Irving-Lewis

Vampire F3s of 614 Sqn taxiing at RAF Llandow, *c.*1952.
614 Sqn archive

Vampire FB5s of 614 Sqn in formation, 1954.
614 Sqn archive

The wreckage of the Avro Tudor airliner *Star Girl* which crashed at Llandow on 12 March 1950. Among those killed was 614 Sqn's Commanding Officer, Sqn Ldr Bill Irving.

South Wales Argus

Standard bearers at the squadron memorial at Pengam. The standard of the left is that of the squadron association, now hanging in St Michael's Church, Michaelston-le-Pit.

614 Sqn archive

Veterans of 614 Squadron Association, led by Association President Glen Pallot, with Parade Marshall, Dai Park behind, march at Pengam for the dedication of the squadron memorial in August 1996.

614 Sqn archive

Members of 614 Sqn outside Cardiff City Hall at the opening of the Wales and the Battle of Britain exhibition in September 2021. From left: AS1 Malembe Makawa, Sgt Andrew 'Shaggy' Oldham, Sqn Ldr Jilly Pritchard, Flt Lt Jude Simpson, and Officer Commanding 614 Sqn, Wg Cdr Olly Walker. Between them all they each wear one of the RAF's different types of No. 1 headgear – including Sqn Ldr Pritchard who wears a Nursing Officer's hat and Sgt Oldham who wears the white hat of an RAF Policeman.

AS1 Anna Harris, Crown Copyright

Members of 614 Sqn, led by the Officer Commanding, Wg Cdr Olly Walker, march past Cardiff City Hall giving the salute at the Wales National Remembrance parade in 2023. Dignitaries included the First Minister, Mark Drakeford, Air Officer Wales, Air Cdre Adrian Williams and Commander 160 Bde, Brigadier Nick Thomas.

Sgt Cathy Sharples

The author, Flt Lt Martin Wade, with a C130 Hercules and aircrew at Great Falls Air Force Base in Montana during an exchange with the 120th Airlift Wing of the Montana Air National Guard in 2018.

Master Sgt Dan Butterfield, 120th Airlift Wing

The badge of 614 Sqn, with dragon and motto – 'Codaf i Geisio' – 'I rise to Search/Seek'

Driver AS1 Terence Cadogn wearing the 614 Sqn insignia, composed from the dragon in its badge to the colour 'flash' worn by its aircraft in the 1950s.

Cpl Toby Grainger

The then OC 614 Sqn, Wg Cdr Phil Sagar, with the Secretary of State for Wales, David Jones MP, at the official opening of the headquarters of 614 Sqn in June 2014.

Ray Edwards, Crown Copyright

of targets aided only by H2S until the eleventh exercise – a simulated night attack – was reached.[39]

Throughout the remainder of the month and into April they trained hard. Having arrived safely at Celone, Tom Scotland and his crew joined the other crews working to bring them to readiness. He recalled, 'All the crews concentrated on a Pathfinder training programme. Navigation trips around southern Italy helped familiarise our crew with Italian conditions and we had more time to perfect our Pathfinder methods and strategy.'[40]

Hugh Fleckney remembered, 'Some fairly intensive training was obviously called for and between then [late March] and the next operation I did, we put in many hours on cross-country flying and bombing runs.'

The work during March and April culminated in a full-scale exercise to test the newly-qualified aircrew.[41]

The practice was held above Taranto – the site of a more celebrated raid four years earlier. The Royal Navy famously decimated the Italian fleet anchored there with a daring strike by torpedo-carrying Swordfish aircraft in November 1940. A rare victory for a Britain which had experienced setback after setback that year. Much had changed since then. By April 1944, the Allies were in a much stronger position than in 1940. At that time Britain and the Commonwealth had faced the onslaught from Nazi Germany and Fascist Italy alone. Although Britain had frustrated German designs in the Battle of Britain, on most fronts they were in retreat. Now, joined by the might of the USA and the Soviet Union, there had been many victories. The Axis had been swept from North Africa, Italy had been invaded by the Allies and had capitulated. Against the Soviets, Germany had faced a disastrous surrender at Stalingrad and defeat at Kursk, the greatest tank battle ever fought as the Red Army turned the tide against the Nazi invader.

The Allies were now in the ascendancy and it was Germany's

turn to retreat. However, the war was not yet won and the final year of fighting would see hundreds of thousands of men and women die, not least in the campaign in Italy. The fighting there was hard and gruelling, and 614 would go on to lose more in this theatre than in all the others combined. However, it was here it would do some of its most vital work, giving support to the bombers who pounded industrial and military targets across the Balkans, northern Italy and the Reich itself. But before it could do that it would have to learn how to mark its targets.

The squadron diary noted that the results were generally good, 'Various mistakes were made but on the whole results were most satisfactory.'[42]

What was the Pathfinder art that they were perfecting? Broadly this is how 614's pathfinding operations would work. Blind Illuminator (BI) aircraft would arrive over the target at 'blitz' minus roughly six minutes to put down flares. These flares lit up the ground so that the Halifaxes designated as Visual Illuminators (VIs), and following close behind, could mark the aiming points (APs).

Target Indicators (TIs) or Target-Marking Bombs (TMBs) were specially adapted 250lb bomb casings containing coloured pyrotechnic 'candles'. At a pre-set height controlled by a barometric fuse, the casing would explode, releasing the 'candles' of predetermined colour which would float down to the ground. They could be set to ignite on ejection to cascade to the ground like fireworks or they could be set to ignite on reaching the ground to form a pool of coloured fire. The radius of ground marking normally produced by a 250lb target indicator was about 100 yards.[43]

The colour used was changed regularly to prevent the enemy from lighting false 'TMBs' safely away from the target, to mislead the main Bomber Force. It was these TMBs which the individual Air Bombers (sometimes referred to as Bomb-Aimers or Bombardiers) in each aircraft sought as they guided their captains into their bombing runs during the 'blitz' period. This could be as brief as three or four minutes or as much as ten to twelve, but it was essential to make it as sudden and brief as possible and get away.[44]

A 614 Squadron pilot at this time, Les Sims gives additional context on how these methods were used and how they could vary. 'Visual and blind illumination were the key ways to illuminate a target by dropping flares by parachutes. This would enable the target-indicating aircraft to go in at a lower altitude to place its target indicators (TIs) on target.'

H2S radar helped aircraft into the target area and, if visual, the bomb-aimer in the nose released. If blind, the H2S operator released. A combination of the two was sometimes used.

'Such illuminations were used to start attacks, but once the smoke and dust created by the bombing started, continued illumination lost its value.

'The aiming point for any main force [bombers] was therefore a coloured target indicator. This was a pyrotechnic of high intensity in varying colours and combinations of colours with barometric fuses which burst the TI at different levels from the ground. Their burning period on the ground was relatively short, and required further placings from the TI aircraft.

'A combination of these two techniques resulted in blind marking by means of sky markers which were coloured flares dropped on parachutes, burning for about three minutes. These could be dropped to burn above, or in thick cloud. Calculations for drift and descent would have been made for the bombs dropped by the main force and aimed at these markers. The techniques of marking with 614 differed a little from those used by UK squadrons,' Les Sims recalled. 'We had no Oboe [a British aerial blind bombing system based on radio transponder technology which was only used in northern Europe by the RAF].'

The H2S radar was key. Its name was a code and has been variously attributed to an abbreviation of 'Home Sweet Home' or the chemical symbol for gas emitted by rotting vegetation. It was sometimes cryptically referred to as 'Y Equipment'.

But it was hard work to operate and to interpret what it told you, Tom Scotland noted. Tricky though it was, he was in awe of its capabilities. 'This brilliant invention enabled us to fly above cloud at night and gain a picture of the earth below us. The picture on the screen generally gave confusing patterns of green textures that had to be interpreted, but seashores and edges of lakes were very clear. Cities were like blobs, but on closer inspection we saw that the blobs had shape. We interpreted the shapes from special maps. Later we were able to enlarge the scale and even identify parts of a city.'[45]

Another vital part of the Pathfinder's equipment was the MK XIV Bomb-Sight. Course, height and airspeed were fed in to the sight from the aircraft's instruments to compensate when dropping bombs. Also, windspeed and direction, as well as the terminal velocity of the bomb were set.

Having completed exercises like the one at Taranto and become familiar with H2S, TIs and Gee navigational aids, crews considered trained were sent on what were called 'squadron private parties' over enemy territory. These were raids over targets not considered to be difficult, but sufficiently well-defended to give crews an idea of what opposition was like. The boss hoped that, 'by the end of the month when all crews had completed the entire course, the squadron would be prepared to carry on in a bigger and better way.'

Raids followed which marked the squadron's transition – from a unit not considered ready, to one that could fly out and face the full force of the enemy and help attacks hit home.

The first came on 10 April when the squadron was called upon again to fly to Bulgaria as target-markers, this time to Plovdiv's marshalling yards. Eight aircraft were sent but they had difficulty identifying the target visually because of ground haze. One aircraft was able to drop its visual markers, but haze prevented observations of bombing, although one crew saw flashes and a glow in the target area.

One aircraft jettisoned bombs and returned early due to bad weather and technical failure. One aircraft burst a tyre on landing, causing its undercarriage to collapse. Because of this, three aircraft had to be diverted. It wasn't spectacular – but it was a reasonable start.

The second raid, however, saw the squadron pay a heavy price when, on 20–21 April, ten Halifaxes were sent to mark and bomb Mestre's marshalling yards near Venice. After joining 614, Tom Scotland had met some of the many Australians who made up the newly-reformed squadron, like Jerry Atwell. He had worked as a barber and played rugby league for North Sydney before enlisting in the Royal Australian Air Force in May 1941.

Jerry joined Tom Scotland for the raid on Mestre. They took off with Scotland's Halifax following in the slipstream of Atwell's bomber.

A threatening weather front sat between them and the target and angry forks of lightning flashed ahead. Clouds thickened and ice began to form on their wings. A message came over the radio – the raid was being recalled. The aircraft turned back and bombs were jettisoned in the sea, but one did not receive the signal.

When the crews got back they learnt that all the other Halifaxes had returned, apart from Jerry Atwell's. They waited for them, but his Halifax did not appear and no trace of it was ever found. 'There were no farewells,' Tom recalled, 'only a cold hard sky offering silent reminder, "This is war!"'[46]

The lost crew of Halifax JN943 were Pilot P/O Gerald Joseph Atwell RAAF; Observer F/S John Malcolm Macleod RNZAF; WOP F/S John Joseph Jones; F/Eng Sgt Will Craig Mckay; Bdr Sgt Victor Barwick Prime and T/Gnr F/S Leonard Henry Hibbins RAAF.[47]

In Tom Scotland's account of the raid, he remembers the target was Hungary – however, the squadron diary records the raid on which Atwell and crew were lost was on Mestre. This is echoed by an account of Atwell's death from the Returned Services League, Australia.

Following a raid on Parma, 614's next Pathfinder mission was on 1 May against Alessandria's marshalling yards, just north of Genoa. The attack showed starkly the risks they would face on every mission they went on.

They had to navigate over thousands of miles – including high mountains, sea and enemy territory to find the target. And they were the first over the target. Flying in a Pathfinder role they had to fly straight and level as they dropped their target indicators precisely on time.

Having awakened the hornets' nest, they had to make good their escape – dodging flak, searchlights and night-fighters. If they survived that they then had to get home, often having sustained damage during the raid. It was no easy task.

The following is the experience of Tom Scotland's crew that night.

'We came in from the south at 17,000 feet with the time 26 minutes past midnight,' Tom recalled. A hot reception awaited them.

German gun batteries, poised on the high ground above Genoa, were waiting for them. They fired, and Tom saw that 'when their venom spat at us, heavy explosions reverberated through our Halifax'. The gunners found their range. 'My stomach tightened with each exploding shell. Minutes ticked by. Nearer and nearer the target we flew.'

Bomb-aimer Ed Riley, lying flat on his front in the nose, guided Tom towards the target. 'Left, left,' he called, then 'R...i...g...h...t, steady, steady, steady.'

'An age passed as Ed held me with the words, "steady, steady". Then, at last he said, "Bombs gone!"'

As noted, it was vital they flew straight and level. All the while, flak gunners on the ground could take aim at the steady target.

'For the next one and a half minutes I continued to keep the Halifax steady, waiting for our camera to take a photograph of the target. The camera flash went off, but so did some very close explosions from the shellfire that had continued to batter us right over the target.'

Suddenly, their Halifax lurched violently. Pieces of shell had hit one of the engines and shrapnel hit cables leading to electronic equipment. The air engineer, Bob Lewis, shouted over the intercom, 'Skipper we've got trouble with our port outer motor. I think we'll have to shut it down!'

Tom prayed that the remaining three engines were undamaged. They had three and a half hours flying ahead of them with high mountains lying across their path, and the weather was worsening.

More warnings came. 'The H2S has gone on the blink,' called bomb-aimer Ed Riley. The radar would have been vital as the weather closed in.

Their homeward route lay southward and Tom nursed the Halifax as they flew into thickening clouds and plunging temperatures. They limped along for an hour and a half. Then navigator 'Hoppy' gave a change of course to cross the Marchigiano Mountains to the east. Their ordeal was about to get worse.

As they flew over the mountains, strong up-draughts and down-draughts jolted the bomber, causing the wings to flex ominously.

'Then,' Tom recalled, 'came the icing.' It began coating their wings, followed by an 'ominous juddering'. The shaking began at the front of the aircraft and built up along the fuselage as the ice overloaded the propellers, causing the whole aircraft to shudder. As more ice formed on the whirling propellers, it became heavier, and so would be flung from them, often striking the aircraft. More ice would coat the propellers and the process would start again.

In the pitch dark, high over the mountains, Scotland could not see the wings but was desperate to know how badly iced they were. 'Engineer,' Tom called anxiously, 'Use your Aldis lamp. Check the wings for how much the ice has built up.'

Bob Lewis used his signalling lamp and pointed its beam along the wings. 'Ice formed very quickly once the air temperature and cloud formation were right,' Tom remembered.

'There's a layer of white ice over our wings and more right along our leading edge,' Bob called.

Tom knew that the ice forming on the Halifax could soon make it unable to fly or cause it to break-up in mid-air.

It was beginning to trap them. To get out of the icing situation, they needed to change height. But they could not get the ice-laden Halifax to climb on three motors and they couldn't descend because of the mountains. 'We had to hope that we could keep flying.'

Scotland was acutely aware that the end seemed to approaching fast. 'Would there be some sort of warning before the wings received too much ice and I started to lose control of the plane?' he asked himself.

He asked Bob to count the thumps per minute of ice coming off the propeller blades and let him know if it increased in frequency. Soon after Bob replied, with grim news, 'There's the same number of ice pieces coming off the propellers as before.' But there was a glimmer of hope. 'I've checked the ice forming on the wings too. It doesn't seem to be increasing.' Tom's spirits rose slightly.

Scotland checked in with their tail gunner, Bill Smith – alone, and at the far end of the freezing Halifax. 'How are you going Smithy?' he called. 'Bloody cold,' replied the luckless 'Tail-end Charlie'. 'There's ice forming around my face and oxygen mask,' he said. Tom tried to comfort him, 'We'll get down into warmer air once we cross these mountains.'

However, the danger mounted. 'Ice is increasing!' called

Bob. Tom increased the power on the three engines, which were already straining. Their speed did not increase as it should have. 'Our ice load had become critical,' Tom recalled, 'accompanied by unbelievable vibration'.

Tom knew he had to take a risk if the faltering bomber was not be torn from the sky by the weight of ice. He decided that, despite the looming mountains, they must descend. He pushed the nose of the Halifax down for more speed and fervently hoped there were no mountains in the way.

'We lost height rapidly. Down, down, down we went.'

'Clouds are thinning out,' called Ed.

Their airspeed increased slowly.

Then, miraculously, the juddering eased. They were flying into air where the ice could thaw. Their engines, which had run so roughly as the ice gathered on the props, 'picked up their smooth song again'.

Despite these good omens, Tom's hands still grasped the control column – his fingers locked in a tight grip. He flexed his hands to ease the pain in his fingers, white from gripping the controls. He remembered feeling ecstatic. They had survived.

Having landed in the wounded bomber, the crew clambered out into the dark. Tom and Bob shone their Aldis lamps onto the battered aircraft. They made their way around the plane looking at the damage caused by the flak. In the quiet early morning of Foggia, Tom recalled the Halifax seemed like some great creature that had brought them safely back. Quietly he said to it, 'Thank you'.

His tent mate, Tony Wheeler, mumbled a few words as Tom came in, and then dozed off. Tom undressed and got into the cold bed. He tried to settle but it was futile. 'The noise and drama of the night's raid caused me to relive the terrifying flight. The firmness and security of my bed seemed an impossible dream.' He tried to sleep but it was hopeless.

*

This would be the lot of 614's aircrews. Nights of terror followed by desperate efforts to rest in a tent on a busy airbase during the day.

But with each mission they would get better at their prime role – marking targets. Arthur Spencer, the experienced Pathfinder and advisor who joined up with 614 in March, told how other developments would help them do their job more effectively from this time onwards. He recalled that by early May 1944 the first air test of the Italian Gee chain had taken place. This would be 'an enormous advance' he believed.[48]

And so 614's new life as a Pathfinder squadron began. As they geared up operationally, so too did conditions at Celone.

There were efforts to protect the squadron from mosquitoes which plagued the Foggia plain. Medical officer, Flt Lt RE Francis, lectured on the dangers of malaria and there was clean-up of the camp. A working party was formed, reinforced by Italian civilian labour, to deal with areas where mosquitoes were likely to breed. Camp inspections were frequent to ensure that all were doing their part in the anti-malaria drive.

Some of the squadron's activities wouldn't need to be under canvas any more once a hut was purchased from a departing squadron. Volunteers put in hours of work to make the hut comfortable, and furniture was fashioned from scrap wood and the interior was decorated. A blackboard and easel with a war map for briefings was put up. It was basic, but an improvement.

Sport was organised and a football pitch was prepared. Intersection games were played and matches with neighbouring units were held. The squadron was able to field 'a very good team'. On Sunday, 9 April, the squadron pitch was used for the RAF v Army match, and an excellent game was enjoyed by the crowd of supporters. The squadron was settling in at its new home.[49]

Those small comforts and morale-boosting sport would be

sorely needed over the following months. From now on 614's range of targets expanded. Oil would be in their sights and they soon began to hit refineries, especially in Romania and Hungary. Intended to stem the flow of fuel for Germany's war machine, the raids would mean long arduous flights. These targets were more heavily defended than railway marshalling yards, ringed by anti-aircraft guns to fill the sky with flak, and the air space around them was patrolled by fighters. All would take a heavy toll on 614 aircraft and crew in the coming months.

CHAPTER 7

Italy, oil (1944)

THE OIL FIELDS and refineries of Ploiesti in Romania produced 30 per cent of Axis oil. This vital target was struck in 1942 by the Soviets and the USAAF. In 1943 'Operation Tidal Wave', an American attack by heavy bombers based in Libya, saw extremely heavy losses.

By April 1944, now within range of airfields in Italy, MASAF launched a sustained campaign against the oil fields. Continuing until 19 August, the RAF carried out 900 sorties against targets, while the American 15th Air Force attacked by day.[1] Around 50 per cent of all Axis oil capacity – in Romania and across Central Europe – lay within range of MASAF's Foggia airfields, and attacks on oil fields continued until the war's end.

The raids were costly. The Germans built formidable defences around the Romanian oil fields, with radar-controlling interceptors from the Luftwaffe and Air Forces of Romania and Bulgaria, anti-aircraft guns and searchlights. Alongside several hundred 128, 105 and 88mm guns, it made one of the heaviest and best-integrated air defence networks in Europe. Also deployed were around 2,000 smoke generators, capable of blanketing the target area in 40 minutes – more than enough time if approaching bombers were detected by radar stations on the Adriatic coast.[2]

Nevertheless, 614's first sorties against Romanian oil fields between 3 and 6 May went well. Although defences over Bucharest's marshalling yards and Steua Româna oil refinery at Câmpina were strong, they marked the targets and all aircraft returned safely. But the next raid would be ill-starred from the beginning.

Seven Halifaxes were to illuminate Bucharest's industrial area on 7 May, but one didn't make it into the air after an escape hatch flew open. Another aircraft's radio failed and it returned to base. The remaining five found and marked the target and although fires and explosions were observed, they still faced a 580-mile journey home.

Returning crews reported flares in the sky following the aircraft. Believed to be so-called 'fighter flares', they were fired by fighters when they spotted a bomber to signal anti-aircraft gunners to stop firing as they moved in for the kill. One Halifax, JP111, was not heard from after take-off and had not returned.[3]

Dudley Egles, the navigator on JP111, recalled a warning from USAAF crews who'd just returned from bombing the same target – albeit in daylight. 'They told us the weather was wonderful, without a cloud in the sky.' They, however, wanted cloud cover as there was a full moon. The squadron's met officer assured them that, at sunset, cloud would build to give them the cover they needed. As they flew through the dusk, Dudley recalled there wasn't a cloud in sight. 'Of course, we pressed on, marked a route point for the bombers following us and marked and bombed our target.'[4] But there was a long way to go before they were safe.

'We had turned on our homeward course when Paddy Beevor, the rear-gunner shook us all by calling over the intercom, "Bandit to port – corkscrew – NOW!" We were at 20,000 feet in bright moonlight and, with a cloudless sky, it seemed almost like daylight.' Norrie Dear, the pilot, threw the Halifax about 'as if it was a Tiger Moth', Egles recalled, but to no avail. A rocket from an attacking Ju88 came through the fuselage into the main plane which immediately caught fire and started to buckle.[5] A report from 205 Group told how a Halifax was seen with one engine on fire pursued by a Ju88.[6]

The bomber began to break up. Egles recalled wrily, 'A Halifax will not fly with one wing so the skipper ordered us to bail out.' Egles was not new to this, having done it two years before. He remembered how cold it'd been when landing behind enemy lines and kept his flying helmet on, tucking the intercom cable into his battledress before leaping into the night.

(More on Dudley Egles and his crew's story can be read in the POWs section.)

Further attacks on the oil fields came in June. There were raids on Giurgiu in Romania, Brod in Yugoslavia, and Budapest's Shell oil refinery on 25 June.

For the raid over the Hungarian capital, eight Halifaxes took off but two returned early with unserviceable guns and navigation equipment. Of the remaining five, three dropped flares but the illumination was off-target. A violent explosion followed by two fires was observed, 'moderate to intense and accurate' anti-aircraft fire was encountered and there were up to 30 searchlights in operation. One aircraft was missing.

Halifax JP107, piloted by W/O Eric Gittings, carried flares and bombs for the mission. There were reports of a four-engined aircraft being tracked by searchlights and being on fire at 00.15 at 10,000 feet in the target area. There were also unconfirmed reports of two parachutes, but at 00.23 some ten to 15 miles south of the target, a Halifax was seen in flames and exploded. This was likely to be JP107. All the crew were killed. They were Pilot W/O Eric George Gittings; Observer F/S Kenneth John Skinner; W/OP F/S Samuel Cater; F/Eng Sgt Alec Shaw; T/Gnr F/S Albert Sidney Harrison Brown and Bomb-Aimer P/O Gordon Melville Williams.

Giurgiu oil refinery in Romania was again targeted on

29 June. Nine aircraft took off but only five released flares and bombs. One Halifax, JN942, piloted by Flt Lt PGG De Bloeme, radioed Amendola at 04.05 reporting its position; then at 04.12 reporting an engine on fire. All contact ceased at 04.22 and the bomber failed to return. The pilot and crew all bailed out and were captured. As well as the pilot, Flt Lt De Bloeme, they were Nav F/O PW Fels; WOP F/O GJ Brigden; F/Eng F/Sgt E Young; T/Gnr F/O HMR Ellison and Air Bomber Sgt WE Way.[7]

July saw the largest number of attacks on oil targets yet with missions on nine separate nights. Targets would be on Romania – Prahova, on 2, 23 and 27 July, and Ploiesti on 26 July; Italy – Fiume, on 19 July; Yugoslavia – Brod, on 8 and 14 July, Smederovo on 16 July and Czechoslovakia – Pardubice on 21 July.[8]

While the attacks on oil fields accelerated, change was afoot for 614. The squadron would be moving from Celone to Stornara – also on the Foggia plain. This was to be the first step in completing the policy of making the squadron controlled directly by 205 Group.

Another landmark was the squadron finally having solid buildings, having lived and worked mostly under canvas since its arrival in North Africa 15 months earlier.[9] The squadron also learnt that CO Wg Cdr WT Russell was leaving the squadron and Wg Cdr JS Laird was appointed as his successor.

Having arrived at Stornara in early May, by 14 July the squadron was on the move again – this time to Amendola, another Foggia airfield just 25 miles away. A 'pleasant domestic site' was found five miles from the aerodrome, about twelve miles from the sea, with much excitement at the prospect of bathing. Just before the move, the cinema gave its first film showing in Italy of the D-Day landings (or *French Invasion* as it was called).

The squadron's review, 'Rise and Shine', gave performances

to packed houses. Football continued until the time of the move, with the squadron possessing 'a very good 1st team and 2nd team'. So serious was football taken that, prior to the move, scouts were sent in advance to look for a site for a pitch at their new home.[10]

Meanwhile, oil targets remained a priority. A raid was planned for 9–10 August, which would see the 205 Group's biggest 'maximum effort' of the campaign against Romania.

The objective was Ploiesti's Romano-Americana refinery. With destruction and damage to other oil plants, the Romana, still in full production, became a most important target. The refinery was producing nine-tenths of Romania's oil and its defences had been greatly strengthened.

The Group would send 81 aircraft, among them eight Halifaxes of 614 Squadron which would mark the target. En route they'd also drop different coloured flares to indicate way-points.[11]

All of 614's aircraft took off from Amendola by 19.40, flying north-east over the Adriatic. The journey to the target, 60 miles north of Bucharest, meant a round trip of 1,100 miles, and the whole journey would take up to seven and a half hours.

As they neared the target, they could be detected by radar stations surrounding the refineries. With experience of battling RAF night raids over Western Europe, Germany had developed an ingenious air defence radar system. Using nine different radar stations, 'Himmelbett' had operators to guide night-fighters to bombers which then could be tracked with their own airborne radar when close enough.[12] They would typically approach from below and behind, knowing the bombers were vulnerable to attack from there.

This is what awaited the Pathfinders of 614 Squadron as they approached over the target just seven minutes ahead of the bombers – due at 23.30. At just after 23.00, Halifax JP259, flown by F/O W Simpson, was nearing the refinery. A Ju88 was spotted closing fast from below on the port quarter. Simpson took frantic evasive action. The night-fighter appeared again, now 400 yards beneath the Halifax. It fired a one-second burst, hitting the bomber's tail-plane. The tail gunner could not fire back as the Ju88 dived to starboard and broke off the attack. The starboard inner engine was hit but kept running. Simpson flew on and levelled the Halifax for the run-in. While he held the big bomber steady, searchlights caught her.

As soon as one found the bomber, others followed – just some of the 40 to 50 searchlights the crew counted that night. They were a sitting target, trapped in the searchlights' cone, but Simpson had to keep her steady and level to ensure their markers were released properly.

Defences were heavy – 'intense, accurate heavy and light AA' they said. They dropped their first flare at 23.24, but the target area was obscured by a smokescreen. Before Simpson could break off, they were hit by flak, damaging the bomb doors and some equipment. He peeled the battered Halifax away and headed back to Amendola. Despite the punishment it had taken, the sturdy bomber brought them home safely.

At 23.15, JP279, flown by P/O Wheeler, started its run-in to the target. Again, they faced a barrage of anti-aircraft fire and, seven minutes later, flak tore into one side of the aircraft, knocking out the port outer engine. To gain height, TMBs were jettisoned. Wheeler took desperate evasive action and decided they could not make a successful run on three engines. The crew reported up to 60 searchlights, which they managed to avoid, and so returned to base.

Halifax JP352, meanwhile, flown by F/O Thomas Sparrow, dropped their flares in one stick. Immediately after they were coned by searchlights twice, and had to take evasive action. Dodging 'very active and accurate defences', they returned safely.

As Halifax JP135, piloted by F/S BM Stephenson, flew towards their dropping point, they fell behind schedule. They had a precise slot allotted to them, timed to the second, but having to dodge flak and searchlights on the approach they got later and later.

After taking hits and being coned by searchlights, they resumed their course, but were coned a second time and continued to be picked up as they flew on. Eventually, they dropped neither flares nor TIs due to their late arrival. As they were leaving the area at 23.55, 30 miles west of the target, their starboard outer engine failed, yet Stephenson managed to nurse the bomber home.

Halifax JP321, flown by W/O Wheldon, was unique in being the only 614 aircraft to return unscathed. Although the target was obscured by a smokescreen, Wheldon's crew found it on radar and made their timed run to target as briefed, dropping their green TMBs and 24 flares in one stick. They returned safely. Two other Halifax crews were not as lucky.[13]

Nothing was heard from either JP282, flown by Flt Lt Langton, or JP110, flown by F/S Caldwell-Wearne. Piecing together other accounts from that night, it's possible to create a picture of what happened to them.

F/S Norman Lord RNZAF, of No.150 Squadron, was flying a Wellington and remembered a night fraught with risk. 'The searchlights were like several rows of a picket fence and we were crazy enough to fly in amongst them in search of our targets – such is the foolishness of youth.'

He saw a Liberator trapped 'by a blue-beamed searchlight' and by several others. 'The bomber became the centre of a concentrated ack-ack attack the like of which I never saw again, or before. They were literally blasted out of the sky in a matter of 30 seconds.'

Night-fighters also set upon JP110. Patrick Macdonald, in his account of RAF raids over Romania, *Through Darkness to Light*, noted, 'It is known that JP110 was hit and set on fire by a Ju88 which continued to sit behind the burning aircraft firing into it. A Wellington apparently dived onto the German aircraft and opened fire from its nose turret. The guns, it seemed, had been previously loaded with day tracer and the ensuing firework display so un-nerved the fighter that it broke away, leaving the Halifax crew to bail out.'

He told how Caldwell-Wearne was about to jump – the last to go of his crew – when the aircraft blew up and he was rendered unconscious. He woke up hanging under his parachute, having no recollection of pulling the rip cord.

It's possible that Norman Lord may have seen the final moments of the other Halifax, JP282, captained by Flt Lt Langton, which crashed at Corbii Mari some 30 miles west of Bucharest.[14]

Additional evidence comes from the POW camp in Bucharest where the NCOs were detained. F/S Norman Foster, RNZAF, of 614 Squadron, who was shot down alongside Dudley Egles on 7 May, was a POW in Bucharest at that time and watched the night's events from the ground. They saw about six planes caught in searchlights. In his diary he wrote, 'One was brought down. Another we saw came right over us and we could see a fighter making passes at him. He finally got away but all hell and fury were let loose on him 'til he did so. On 10 August we could see smoke rising from Ploiesti and, by dusk, the whole sky was covered with a smoky haze. We heard from Langton's Wireless Operator [W/O HG Poynton] that his crew were blown up – he was the only one to get out. So there goes Bob Langton, "Will" Martyn and Nobby Clarke.'

He would also eventually meet some crew members of JP110. Between 11 and 17 August, Foster went on to note, 'Caldwell-Wearne's crew came in the other day. They all got out but C-W himself has been sent to the schoolhouse [the officers' camp]. Gibson, his Air-Bomber, is with us now.'[15]

All but one of the crew of JP282 died after their Halifax was shot down. The crew were Captain Flt Lt Robert Douglas Langton; Navigator F/O Thomas Joseph Rogerson Turner; Air Bomber P/O Walter Joseph Martyn; F/Eng F/S Herbert

George Jones; Air-Gunner F/S John William Clarke. Another name not listed in the squadron diary is Air Bomber F/S Peter Brunton. The sole survivor was W/O HG Poynton, who F/S Norman Foster had heard from at the POW camp in Bucharest.[16]

The only crew member to die when Halifax JP110 was shot down was Navigator W/O Harry Bath. Although interviews with the crew following the end of their captivity told they believed he'd escaped. as his parachute was found in good condition,[17] he was buried in Bucharest's war cemetery alongside those killed when JP282 crashed.

The remaining crew members, Pilot F/S Claude Caldwell-Wearne; WOP F/S HG Jones; F/Eng Sgt D Sambrook; T/Gnr Sgt RL Cote and Bomb-aimer/Air-Bomber F/S RB Collins, all parachuted safely and became POWs.

Of the eight 614 Squadron aircraft that flew to Ploiesti that night, two did not return, but of those that did, one by rights should not have, and most of her crew did not. But that night had one more story to tell for 614 – and it would be a story of miraculous survival.

One of the eight Pathfinder crews marking the target was that of Canadian F/O Bruce Prange. The 21-year-old office worker from Kitchener, Ontario, had joined 614 in May 1944, and completed eleven missions.

Flying in Halifax JP225, Prange's crew couldn't identify the target due to the smokescreen, despite making two runs. So Bruce decided not to release his red TIs. At 00.30 he headed for home and was around 175 miles away from Ploiesti, just south of the Danube, when there was a blinding flash under the aircraft's nose. They'd been hit by anti-aircraft fire, followed by another explosion in the bomb bay as one of the red TIs detonated. Fire broke out. It seemed the bomber was finished. Bruce ordered the crew to 'put on parachutes'. Then the intercom went dead.

In the confusion Bomb-Aimer W/O Allan Fox, also a

Canadian, accidentally pulled his parachute chord and clambered aft to get another. As he passed the pilot, Prange saw that Fox was covered in blood, so told him not to bail out and reassured him that he hoped to soon have the bomber under control.

Despite his injuries, Fox grabbed a fire extinguisher and put out the fire in the bomb bay before returning to the nose where he found another fire near the oxygen supply. If that took hold, they'd be finished. Again, he struggled aft, this time searching amid darkness and smoke for the oxygen controls which he found and turned off. The badly wounded Canadian then adjusted the fuel cocks to pump petrol to different wing tanks to keep the aircraft stable. The aircraft was safely under control – for now.

Clambering down the fuselage, the bloodied Fox searched for his fellow crew members and found they'd gone. In the confusion of fire and smoke the four had bailed out from what seemed a doomed aircraft. Because the intercom went dead, the order to bail out wouldn't come, so they clearly decided not to wait for one. Prange never gave such an instruction – only ordering them to 'put on parachutes'.

Fox gave Prange the news. The wound to Fox's arm was bleeding profusely and he gave bandages to Prange for him to dress the wound. Bruce couldn't do this and control the aircraft, so Fox applied a tourniquet himself and, as he lost more and more blood, lay down in the rest position.

For four hours, the young pilot nursed them home on the last course given by the navigator. With the hydraulics not working, Prange could not close the bomb doors and the glazed nose was smashed by hits from the flak. The icy wind howled in. 'It was bitter cold at 10,000 feet,' he recalled.

Coming in to land, Prange knew that he would not be able to circuit the airfield. His radio was gone and their flare pistol was lost – giving no way to signal to the control tower. He had one chance to get the Halifax down and, despite the red flare fired from the tower to deny permission to land, he came straight in and touched down successfully.[18]

W/O Fox later said, 'It was a miraculous piece of flying. Bruce is a wizard pilot and I don't know how he did it.' Prange recalled that he did not feel fear until after landing. After he had safely coaxed the bomber down and brought it to a halt, he started to shake and could not sleep for 24 hours.[19]

The four who bailed over Romania became POWs. They were F/Eng Sgt J Forshaw; Observer F/S GSA Squire; WOP/AG F/S EMP Steel and T/Gnr Sgt DH Jamison.

The Group admitted it was a 'most costly operation'. The 81 bombers of 205 Group delivered over 120 tons of bombs that night, but by dawn the next day 32 aircrew were dead or dying, 22 more were POWs, and 13 bombers were lost, including the two from 614.

It added, 'Ground defences, estimated at 100 searchlights and 60 heavy guns, concentrated on the illuminating Halifaxes [from 614 Squadron] and they were repeatedly coned with the result that, in addition to two which are missing, two more were prevented from making bombing runs because of the need for constant evasive action, one could not open the bomb-doors through flak damage and one was late owing to evasive action over Bucharest.'

Fighter opposition and flak was fierce too. 'Ju88, Me109, Me110 and FW190s were identified and persistent attacks were made with tracer and rockets.'[20]

There was recognition for 614 crew members on that desperate night. On 20 October, F/O Prange was awarded the Distinguished Flying Cross for the bravery he showed.

Prange's citation praised his 'iron determination and great skill' in flying his severely damaged aircraft to base and making a safe landing, adding, 'He displayed great courage and devotion to duty in most trying circumstances.'

On 21 November, W/O1 Bernard Allan Marvin Fox was similarly honoured. The 'efficient, conscientious and resourceful' Fox was praised for extinguishing fires in the aircraft despite suffering from numerous injuries caused by shell splinters. 'Owing to his courage and determination, W/O Fox enabled his captain to fly his aircraft safely back to base.'[21]

Following the costly raid of 9 August, oil targets continued to be the focus for much of August. As the Red Army advanced on Romania, so the attacks were stepped up. On the nights of 17, 18 and 19 August, raids on oil installations dropped almost 3,000 tons of bombs.[22]

Again, 614 would be pitched into the battle, and again they would be grievously wounded. By 17 August, they were back over Ploiesti again, this time to mark and bomb the Xenia oil refinery.

Orders from 205 Group said, 'The enemy had commenced to repair Romanian oil refineries by "cannibalizing" damaged installations. It was, therefore, important to destroy all undamaged equipment in refineries.'

To overcome the heavily-defended target, the maximum number of bombers – 63 Wellingtons and nine Liberators – would be over the refinery in the shortest time to swamp defences. Leaving their Foggia airfields between 19.10 and 19.43, the slower Wellingtons left first. The six Halifax Pathfinders would mark at 23.00 with the last bomb dropped by 23.15.

One of 614's Halifaxes turned back with engine trouble. The remaining five dropped 20 green TIs, but smoke soon obscured the markers. A violent explosion and large fire was caused with the first TIs, and several other sticks of bombs were seen to burst around the markers.[23]

Many had narrow escapes. Flt Lt Sid Brownless of 614 Squadron, who flew in F/O Charles Symonds' Halifax JP227, recalled, 'On our first run over the target we had very heavy flak and broke off the run to try again. This time we were caught and coned. The flak was behind us but very close and we had some scattered hits – we thought we were finished.'

Another 614 Squadron pilot, Les Sims, explained how Pathfinders, as the ones first over the target, would run the gauntlet of anti-aircraft guns who'd find their aim.

'The first shot was always well behind and it took around three minutes before successive shots caught you. With tail gunners reporting the position of each burst, we went in straight and level on auto-pilot, correcting course on guidance from the bomb set [operator], so dropping our illuminators or TIs with precision from a steady height and a fairly steady aircraft – subject to flak and night-fighters. Always we sweated during the last 30 seconds or so, as the bursts to the rear got nearer.'

Having just out-run the flak, Sid Brownless tried to drop his markers. 'By this time the enemy had the target smokescreened, and on our third run-in I dropped four green TIs and three 1,000lb bombs using the H2S radar. This was our first time to use this method and we were quite successful.[24]

'Hugh Fleckney of 614 Squadron was flying Halifax JP232 and was caught by a Ju88 from starboard on the run-in. Two other enemy aircraft opened fire at the same time. He corkscrewed and the bomber was not hit.'[25]

He recalled that, unusually, they had a mid-upper gunner (MUG).

'Running in to the target we were in a fairly high pitch of apprehension. Suddenly, the MUG shouted, "Fighter, fighter, starboard – go!" and I pulled a hard bank and rudder to starboard and down. As we went down to the right the MUG opened fire and I could see his tracer arcing out towards one of the decoys. The tracer was falling well short of the enemy aircraft. I told him to cease firing and, simultaneously, our two starboard propellers ran away. By heaving on the stick and pointing the nose skyward we were able to bring them back under control and we resumed our run-in to the target.

'Soon after, the MUG again shouted over the intercom, "Fighter, fighter, port – go!" I corkscrewed to port, hoping like hell that the propellers would not repeat their performance. Then rear-gunner, Dickie Billen, shouted "Fighter, fighter, starboard – GO!" and immediately opened fire on a Ju88. By this time, Fleckney said, "I was a bit hyper-active and I reefed the aircraft into a tight starboard turn, such that Dickie couldn't keep the guns on the Ju88."'

Happily, neither could the fighter, apart from a 20mm shell which hit just below the rear turret but did not explode. 'We emerged unscathed if thoroughly terrified.'[26] All 614 aircraft returned safely from the raid, despite the attentions of the night-fighters.

That night, alongside 86 tons of bombs, 845,000 'nickels' (leaflets) were dropped. Used to sow doubt and discord, the leaflets appealed to Romanians to destroy the 'fog' defence system and to understand the 'fog of German lies' keeping Romania in the war. The translated text read:

ROMANIA IN FOG

As you have seen, the German fog cannot prevent the destruction of the oil refineries at Ploiesti

BUT

ROMANIAN LIVES CAN STILL BE SAVED, IF YOU KNOW HOW TO IMPOSE AN END TO THIS KIND OF CAMOUFLAGE.

Also the fog of the German lies, made up specially to keep Romania at war, cannot prevent the United Nations smashing any resistance on Romanian land.

BUT

Hundreds of thousands of Romanian lives can still be saved, if you can impose the ending of the policy of sacrifice of Romania for Germany.

THE FATE OF YOUR COUNTRY LIES IN YOUR HANDS. REMOVE THE GERMAN FOG.

Dropped by the British Royal Air Force and the United States Army Air Force.[27]

The leaflets give a sense of how strong the pressure – from Allied attacks and advances – were on Romania by this stage in the war.

That pressure told on 23 August when King Michael of Romania and his supporters arrested Marshal Antonescu and deposed his pro-Fascist government. That night, Romania's armed forces defected from the Axis. A week later the Soviets entered Bucharest.[28] It would be a decisive moment for POWs held by Romania, and meant their freedom would soon come.

Despite these dramatic events in Romania, there were still other sources of Axis fuel. The next raid on an oil target would take 614 over Hungary.[29]

Part of a force of 81 aircraft, seven of its Halifaxes were to illuminate the Szőny oil refinery in Hungary on the night of 21 August. A modern facility on the Danube, 45 miles from Budapest, it was severely damaged on 16 June, but repaired and now produced 5,000 tons of oil per month.[30]

Defences were intense, with heavy AA scattered en route near Lake Balaton. Halifaxes made two runs over the target and dropped TMBs and flares. Halifax JP275 was hit by flak, causing fires in the bomb bay which were extinguished and the aircraft limped home. The crew of JP223, flown by P/O Thomas 'Barney' Sparrow, were not so lucky.[31]

F/Eng Sgt Geoff Curtis had only just joined its crew, replacing an engineer who was sick. 'Perhaps,' Geoff mused, 'he knew something that I didn't.' He joined the briefing and met the crew. The flight to the target was uneventful but, on the run-in, things began to liven up.[32] Between Györ and Lake Balaton, tracer fire lurched up at them and night-fighter flares were seen.[33]

Geoff remembered heavy flak causing problems for their flare pattern. 'Having completed our runs and having laid our markers, we turned out of the target area which was, by then, well alight.'

As they set course for home, the skipper said, 'Everyone

OK, bomb bays empty?' before calling to Geoff, 'Engineer change tanks – how much fuel left?' Curtis clambered aft to midway up the fuselage where the fuel cocks were. He had just got back to his position when the intercom crackled 'Night-fighters – CORKSCREW!'[34]

'The plane screamed around the night skies and cannon shells exploded in my position, blowing up the oxygen system beneath the floor and flames erupted at my feet. I felt a searing, jarring pain in my right hand and leg.' The Halifax looped over on her back and then, with engines roaring, dived earthward.

'The "G" effect was so great that the 'chute seemed to weigh a ton. The aircraft flipped over again and I was deposited in the nose by the front exit hatch in time to see another crew member lift off the hatch door and make ready to exit.

'Unfortunately, he was snagged by some projection, so I put my foot behind him and heaved.' Out he went, followed closely by the wireless operator. 'I'd already ascertained the pilot was beyond help so I seated myself on the floor of the aircraft with my feet dangling through the escape hatch and levered myself into the air.'[35]

All but three were killed when JP223 was lost. The dead were Pilot W/O Thomas Bernard 'Barney' Sparrow; T/Gnr F/S Douglas James Heath; Mid Upper Gnr F/S Norman Stuart Yates and Bomb-aimers F/S John Graham Evans and F/O Samuel Kwinter.

F/S G Curtis; WOP/AG F/S MR Patriarca and Navigator F/S Clifford Harold Sparks bailed out and became POWs. (More can be found out about them in the POW section.)

A sad postscript to 614's operations in August 1944 was the loss of another crew, not to enemy fire but while training. Halifax JN912 was returning on 14 August when an engine caught fire and it had to ditch. As it hit the water the aircraft turned on its back and the dinghy failed to inflate.

Tragically, all but two of the crew were lost with the

sinking aircraft. The two survivors were P/O JE Homes and LAC H Isaac. In the water for 19 hours, LAC Isaac struck out for land to fetch help. He raised the alarm and a party of Italian Marines went to rescue P/O Homes. They reached the exhausted survivor, but tragically, such was his state that he died almost immediately.[36]

The names of the victims were Pilot F/O Francis Andrew Sanderson, RNZAF; WOP F/O Joseph Edward Albert Homes; F/Eng Sgt John Hume Anderson; T/Gnr Sgt Harry William James Mitchell; Nav F/S Norman Ernest Hodges; Bomb-Aimer F/S Arthur Alexander O'Connor. There were passengers too: Flt Lt Albert George Edwards; Cpl Harry Newstead and Lt John Leonard Caro, SAAF.[37]

The Romanian campaign was at an end, although raids on oil targets would continue. This costly campaign was as vital as any the RAF fought, yet it remains obscure. Sgt Gordon Cormie, of 40 Squadron, part of 205 Group, ruefully told of how it's viewed. 'People often ask me what I did during the war and what happened. When I've told them that I was a prisoner of war, they have turned and said, "Oh, where?" but when I add, "I was in Romania", that's the end of the conversation. Romania did not conjure up any idea of war. The big raids on Berlin, Hamburg and so on got the glory. What we did seems sort of incidental. I assured such people that a bullet is a bullet and a shell is a shell whether it's over Berlin or over Bucharest.'[38]

CHAPTER 8

Italy, POWs (1944–1945)

THOUSANDS OF THE RAF personnel were shot down over enemy territory in the Second World War to then spend the rest of the war behind barbed wire. Many are familiar with the fate of POWs in German camps, but less well-known are the stories of those captured in Romania or Hungary, as 614's downed airmen were.

These nations' alliances with Germany would have a profound effect on the fate of 614's POWs. For some it would mean freedom before war's end, for others it would mean misery and coming face to face with death.

We left F/Eng Sgt Geoff Curtis as he leapt from the doomed Halifax JP223 following the attack on Szőny oil refinery, and we rejoin him as he is plummeting to the ground.

'Whilst falling from the plane, I failed initially to open my parachute due to my wounded hand. I tugged at the rip cord but it came apart due to the damage done to the pack as a result of the fire – and all the time I was plummeting earthwards.

'In my mind's eye I could see the face of my safety equipment officer as he said, "Remember, count to ten to ensure you are well clear of the aircraft."'

Geoff managed to deploy his parachute and reckoned he

was at 800 feet. Falling further, he crashed through trees, feeling he'd 'bounced off every branch'.

'I hit the ground about 300 yards from our crashed aircraft. Ammunition exploded from the wreckage and bullets whizzed over my head. I lay where I had fallen, numb and bewildered.'

It was some time before he could collect his thoughts. As the explosions died down he decided it was time to move.[1]

Another raid over Hungary resulted in capture. This time, against marshalling yards in Szombathely on 22 November. Eight aircraft from 614 took part and faced substantial opposition. Two Ju88s and several unidentified aircraft were reported, one attacking with rockets. One Halifax JP321, flown by F/O H Wheldon, was badly mauled. Twice attacked by a Ju88, cannon shells set fire to ammunition boxes in the rear fuselage, causing the rounds to detonate. F/Eng Sgt K Briggs extinguished the flames and the Halifax was brought back to base. RAF photographs showing the bomber with a gaping hole in the fuselage floor and peppered with smaller holes incorrectly state flak as the cause. For this action Wheldon later received a DFC and Briggs the DFM.

All except one returned from the raid but Halifax JP323 was missing. Tail gunner F/S Roy Bennett recalled this was the crew's last operation of their tour. But omens were bad. Lining up to take off, a fuel pipe burst and they had to change aircraft. They were first attacked 30 miles from the target, Roy recalled. 'The plane shuddered – I yelled down my intercom – violent weaving and the skipper really threw that plane around.' Having dropped their TMBs, they were attacked again. 'Bullets came into the plane, tearing her to pieces. She staggered like a half-knocked-out boxer and both starboard engines burst into flames.' The pilot, P/O John Nicholson, then ordered, 'Prepare to abandon aircraft.' Roy sat on the edge of a rear escape hatch and watched the starboard wing fall away, then leapt from the doomed bomber. All the crew, apart from the pilot, bailed out.[2]

WOP F/S Syd Barlow from Wigan remembered his escape. 'We'd had our harness adjusted back in England over our full flying kit' (this was underclothing, battledress, Irvin flying jacket and more). In warmer Italy they never used the outer flying suits and he'd never had his harness readjusted.

'In consequence, all my clothes were pulled over my head. Looking up I could see the "V" of my battledress about twelve inches above me. The only way I could look down was by grasping the parachute straps and literally pulling myself out of my clothes.'

He thought he saw another parachute, but it was his own reflection in a small lake he was fast approaching.

'I'd always been a rotten swimmer and dreaded coming down in the sea.' Bracing for impact, he looked down. 'My reflection was almost on me when I pulled the life-jacket toggle to inflate it.' Hitting the water, his heart hammered and he spluttered desperately. Furiously kicking his legs he was surprised to touch solid surface and found that he had, in fact, landed in two feet of water. Wading to its side with his flying boots full of water, he dragged his parachute behind him.[3]

In the previous chapter we left Dudley Egles bailing out from his stricken Halifax. Having heard tales of aircrew being shot at while falling, he made sure his drop was as long as he dared and counted to 40 before pulling the rip cord. Below him, he could see the Danube shining in the moonlight. 'I could not tell whether I was going to land in it, south of it in Bulgaria or north of it in Romania.' The wind decided his destination – Romania.[4]

After landing, thoughts turned to avoiding capture. Dudley hid his parachute in some bushes and walked westwards where he believed Yugoslavia lay. As an experienced navigator, he knew how to find his way by the stars, which were still visible. He'd walked for a couple of hours when he saw some farm buildings. Carefully circumnavigating them, he decided to rest in a haystack. Exhausted, he fell asleep, but didn't rest for long. 'The next thing I remember was waking up surrounded by dogs barking and peasants waving pitchforks. I tried to explain who I was, to no avail, and they "escorted" me back to the farmhouse.'[5]

After *his* landing, Syd Barlow was not alone for long. 'Suddenly, I became aware of someone in the half-light looking at me.' Unnerved, he said the first thing that came into his head. 'What's cooking?' he asked.

'Is that you Syd?' a voice asked. With some relief, Syd realised it was Navigator F/S Bob Hitchen. He'd landed a minute before Syd, and watched him come down. He'd lost one of his flying boots.

The resourceful Syd took out the pocket knife he always carried on operations and fashioned a spare boot from pads in his Mae West and parachute cords. The pair decided to rest until daylight in a small hollow protected by trees. Having settled, they opened their standard-issue escape packs, but were unimpressed. Each pack was supposed to have a map printed on silk of the region they were in, but they found instead theirs were of France and the Low Countries. There was money to spend should they need it, unfortunately it was in US dollars which would be useless. The rest of the contents were similarly uninspiring. 'There were a few Horlicks tablets, water-purifying tablets, a folding razor and a piece of soap. Not a lot to keep us going.'

They were freezing cold and huddled together for warmth, sleeping on and off until daybreak. They decided, like Dudley Egles, that they'd head towards Yugoslavia. They were an 'odd-looking pair' he remembered. 'Bob with a gouty foot and one boot, me, similarly attired but limping painfully as huge blisters had developed on the backs of my ankles.'

They made a plan and decided they'd walk along small country roads, keeping to the nearest compass direction and dodging behind trees or in ditches when they heard signs of anyone approaching. They continued in this fashion for a few hours until they came to a blind corner and saw a horse and cart carrying two farmers drive out of a field. They were, recalled Syd, 'rather surprised to see us'.

He made a desperate bid to communicate, which was doomed to failure and quite understandable, seeing as he came from the only Welsh squadron in the RAF. To the farmers' inquiries in Hungarian, Syd cheerfully replied 'Bore Da' ('Good morning' in Welsh). His uncle was a Welsh speaker and had taught some to Syd. He vainly hoped that it sounded a bit like Hungarian. With a wave they continued on their way, but hadn't gone very far when they heard the sound of horses' hooves pulling a cart clattering up the road behind them. Their plan of escape to Yugoslavia quickly began to unravel.

The farmers had raised the alarm, and in the cart was a pair from the local militia. Syd and Bob faced a decision. Could they overpower the two Home Guards and carry on, or would they surrender quietly? The Home Guards gestured to the airmen to climb aboard the wagon. Syd recalled, 'We decided to accept. We were hungry, frozen stiff and in considerable pain from our feet.'

Looking back, Syd admitted he'd never been proud of the fact they went with them. He said, 'When you read stories about escape, there are reasons for success or failure.'

The evader had to have knowledge of the language; there must be a friendly population prepared to help and the evader should have a disguised appearance.

'As none of these categories applied to us, we gave up. The brutal truth was that we must have stood out like a sore thumb in our disheveled appearance.'

The airmen were taken to a village where people surrounded the wagon in awe. They looked up at them, Syd remembered, 'as though we were Martians'.

As they passed one house, a lady went in and returned

with a white jug of steaming broth and offered it to them. The smell was 'delicious' Syd recalled, 'I was just going to take it from her hand, when a man gave what I took to be an "oath" and knocked it out of her hand, sending the jug and soup flying.'

They were taken to a drill hall in the nearby town of Zalaegerszeg. There, they met a Hungarian colonel, who was in charge of 20 soldiers.[6]

*

Having been taken to a farmhouse, Dudley Egles found the farmer had a smattering of French and they could talk.

They had all thought he was German, as the Luftwaffe uniform was a similar colour to his RAF blue battledress. He pointed out that he was a British officer and the farmer said that he'd have to take him to the local village. He gave Egles some bread and 'apa dulce', literally sweet water, and drove him to the village where he handed him over to the police.

Dudley recalled, 'I was of great interest to the locals – the first RAF officer they'd seen. I was "introduced" to the mayor and daughter who also spoke French and asked me if there was anything I needed.' Egles asked for some soap. 'She went away and returned with a tiny sliver. It was only much later that I realised what a sacrifice this had been.' Spending the night in the local jail, he was collected by an army truck the next day with other members of his crew and some Americans.

One of his crew was the Rear-Gunner Sgt Paddy Beevor. Dudley recalled he was 'somewhat disgruntled, rather bruised' and, like Bob Hitchen, only had one flying boot. The other had got trapped in the tail turret of the crashing bomber as he bailed out. He was picked up by a Romanian army patrol who laughed at his lack of footwear.

'We were told we were being taken to Bucharest and then be sent to prison camp.' They were first taken to an army camp where he was interrogated by an intrigued Romanian general. He demanded to know why they had 'killed so many civilians' with their bombing? Dudley tried to explain, in his rudimentary French, that in war it was almost inevitable that there would be civilian casualties and said many English civilians had suffered bombing.

Then, his schoolboy French let him down. 'I wanted to tell him that, as he was a fighting man, he should understand this and I said, "Vous êtes un soldat." He then went almost blue in the face and shouted, "Je ne suis pas un soldat! Je suis un officier! Je suis un general!" and had me wheeled out.'

They were then interrogated individually by 'Gestapo types' who threatened all sorts of dire consequences if they did not tell them their squadron, aircraft type, base aerodrome, etc. Dudley remembered, 'I only gave my name, rank and number.' He was interrogated by two Luftwaffe officers who spoke excellent English, having gone to university in Britain. While being searched they found the small hacksaw blade in the hem of Dudley's battledress but didn't find the compass hidden beneath his Observer's brevet.[7]

*

Geoff Curtis also entertained thoughts of escape via Yugoslavia. Wandering, dazed, he remembered that the target was close to the border and so hoped he was already in the country and could meet up with partisans. Injured after falling through branches as he descended, he staggered through the dark and eventually came to a farmhouse. He knocked on the door. A woman carrying an oil lamp opened the door. Geoff said, 'Comrade, Yugoslavia? Me, English.' The woman backed away to the shadows, which, he recalled, was not a surprise considering the state he was in. A man in uniform appeared. Exhausted and covered in blood, Geoff was in no position to fight and the man grabbed him and pulled him into the house. He spoke some English and began to question him, but stopped to ask Geoff if he had a medical kit. When he said he did, he took it and began to dress Geoff's wounds. One of the airman's fingers, hanging limply – 'a mass of pulped flesh' – was splinted. Nothing could be done for his leg wound, however.

In daylight, Hungarian soldiers arrived and he was dragged out of the house and put up against a wall. He shivered in the cold and thought they might think that he was afraid. A soldier trained a machine-gun on him. Geoff thought 'this is it'. He remembered mentally urging them to get it over with if they were going to do it, but suddenly there was a disturbance and some soldiers appeared with what looked like another airman. It was F/S Romes Patriarca. He didn't have the chance to speak to the wireless operator, known as 'Ron', and he was taken away. At least he now knew Patriarca was alive.

Geoff was taken to a barracks. A soldier gave him some food which the starving airman wolfed down. The guard also told him he believed three of the crew had survived. The next day, Geoff was taken out of the cell and met Ron again. He told Geoff how he'd been taken to the crash site just after he was captured and saw the grisly sight of the bodies of the pilot, navigator and gunner still in the aircraft. It made him physically sick, he told Geoff. They were later put on a train and told they were going to Budapest. It would be the start of nine months of gruelling captivity which would nearly kill them.

Geoff and Ron were taken to a hospital in the Hungarian capital. Geoff had many injuries after his escape from the doomed aircraft. He had ruptured his right kneecap as he landed, broken his finger and gashed his leg. Ron also needed urgent medical attention. Like Syd Barlow, his parachute harness had been fitted too loosely and, as he descended, the buckle and the harness battered his midriff. Geoff recalled that Ron's abdomen was 'black and blue', nearly rupturing his appendix and leaving him in great pain. He would later have to have his appendix removed. At this point, Geoff and Ron were separated. Geoff was taken to a prison elsewhere in Hungary.

After he and Geoff parted, Ron was taken to hospital and was later moved to a Budapest jail where he was held in solitary confinement for three weeks and beaten and interrogated. He recalled the kindness of Jewish workers who found cigarette stubs for him to smoke.

The relatives of all those missing would be contacted by the squadron. Here is the telegram sent to F/S Patriarca's family from the CO Wg Cdr Laird, three days after the raid:

> 24th August, 1944.
> Dear Mr Patriarca,
> It is with deep regret that I have to confirm the sad news that your son 1390889 Flight Sergeant Patriarca, R. failed to return from operations against the enemy on the night of 21st August, 1944, and at the same time to give you as much information as possible about the night in question. The squadron was operating against a target at SZŐNY in Hungary and your son, flying as wireless operator with his usual crew, was among those taking part in the raid.
>
> All the aircraft were safely airborne and no messages or distress signals of any description were received during the operation. Your son's aircraft did not return to base and nothing has been heard since.
>
> Reports were received from other crews that they had seen an aircraft hit by flak and it is possible, but not definite, that this might have been your son's aircraft. I naturally do not wish to raise your hopes unduly but it is possible that the crew may have baled out successfully or made a forced landing and that some good news may yet be heard. You can rely on being informed immediately if this is so.
>
> In this connection I must ask you not to make any announcement to the press about Ron being missing. Such

announcement, if it fell into enemy hands, would definitely jeopardise the crew's chances of making their way back to neutral or friendly territory.

Flight Sergeant Patriarca had been with the squadron for some time now and was classed as one of our most promising wireless operators.

Both he and his crew were very popular with us all and we feel their loss very badly. All the squadron wish to join me in offering you our deepest sympathy during this anxious time and we all sincerely hope that some good news will yet be heard.

(JS LAIRD) WING COMMANDER [614 Squadron]

L Patriarca Esq. 36, Sandbourne, Brockley, London, S.E.4.[8]

Geoff received hopeful news. A guard who spoke English told him he'd be kept in the country and not be sent to a German camp. He was then moved to the Cegléd area where he met other prisoners – Americans, British, Australian and a South African. Hopes were dashed when they were marched to the town of Komárno and taken by train across the frontier and into the Reich.[9]

The day after the raid in which Curtis and Patriarca were shot down, another raid over Hungary saw more loss. Halifax JP109 was one of five from 614 Squadron marking the marshalling yards at Miskolc.

Navigator W/O Mervyn Bradford recalled he and the crew owed their lives to the level-headed pilot, Flt Lt Walter Carrington. Over Hungary a night-fighter fired two rockets which struck their port wing. Fuel tanks erupted in flames and the Halifax went into a corkscrew dive. Bradford remembered that Carrington stayed at his controls to stabilise the bomber and allow the crew to escape. Had he not done so, Mervyn recalled, they all would have died.

Of the crew, only Walter was badly injured. As he was the last to jump he was caught in the bomber's slipstream and landed close to the Halifax as it crashed. Carrington lay twisted on the ground. One of his eyes had come out of its socket and hung on his cheek. His jaw was broken and he had a gash to his head. Hungarian soldiers appeared and told the others to leave him as he was dead, whereupon Walter opened his good eye and showed them he was alive. The impact of landing had also shattered his pelvis and the badly injured pilot was separated from the rest of the crew and taken to hospital. Perhaps uniquely among 614's Hungarian prisoners, he avoided being taken to Germany and escaped with the help of a Jesuit priest when the Russians captured the area a few months later. After the war Walter would become internationally renowned as an exponent of the Alexander Technique and inspired generations of teachers.[10]

*

Roy Bennet had landed flat on his face in a muddy field but was uninjured. Burying his equipment, including his .38 revolver, as he 'couldn't hit a barn door with it', he was approached by an old man who took him to his farmhouse. The next day, he was joined by flight engineer John Wiles. But, soon after came Hungarian soldiers who chained them up and took them to the aircraft crash site. They saw their dead pilot, P/O John Nicholson, 'a sight', said Ron, 'that will always stay in my memory'.

While standing at the wreckage they were offered some wine by a young Hungarian girl. They gestured to their guards to unchain their hands so that they could drink, which they did. The guards became distracted by the girl, and the pair sidled away. But escape was short-lived.

'Three hours later, walking through a small village we turned a corner and ran smack into the SS who, with drawn revolvers, made short work of our freedom.' They were taken to the same village hall as Syd and Bob, where the final survivor, Bomb-Aimer F/S Ron Cartwright, was also taken.[11]

The other three were saddened to learn of Nick's death. 'It was just like losing a brother and it has remained so ever since,' Syd recalled. Next day they had visitors – eight Wehrmacht soldiers, including an NCO and an officer who Syd recalled 'looked a mean sod'. He recalled the German officer and the Hungarian colonel were soon in heated conversation, which seemed to be about their fate. 'They shouted and bawled at each other, whilst we didn't have a clue what they were saying, only that we were the subject of dispute as they frequently pointed over to us.' It was, Syd said, 'a bit like tennis at Wimbledon, as we jerked our heads from one protagonist to the other, hoping our side (the Hungarian) would win. In the end, however, the Hungarian shrugged his shoulders and the German motioned us to stand up.'

They were marched to a courtyard flanked by a nine-foot wall. Placed with their backs to the far wall, the six soldiers faced the airmen about 15 feet away. 'The officer shouted a command and the soldiers came up from the stand-easy position, and pointed their rifles at us.' Syd recalled, 'I couldn't take this in. We were all numb.'

He then gave another command and the rifles were brought down. He turned on his heel and went back into the barracks, leaving the soldiers in their original position.

The officer soon returned with, Syd remembered, 'both hands behind his back as if he was concealing something'. He assumed it was the white cloth which, when dropped, is the signal for the squad to fire.

The officer then shouted a further order, and the rifles were raised again. Syd recalled grimly, 'I'll always remember Ron Cartwright shouting, "Bugger 'em, don't let them see you're scared, mates!" I tried to make a joke of it and said, "They'll not need to, they'll only need to look down at our legs!"'

The officer shouted another order. 'He whipped his hands from behind his back producing a camera and, all in one movement, took our photographs,' Syd recalled. 'He probably wanted to show terrified British airmen. He succeeded, too.'

Syd confessed he'd never know what led to that terrifying moment. 'Was it an elaborate charade or did that Hungarian Colonel really save our lives?'[12]

For all the 614 prisoners, arrival at their camps meant the reality of life behind wire began to sink in.

In Romania, Dudley Egles and the other prisoners were paraded and marched through the city to their camp – a disused high school with windows covered in barbed wire. A day later the NCOs were sent to the Mihai Viteazu Barracks and Dudley was separated from his crew, who were all sergeants. A resourceful crewmate had a parting gift for Dudley. 'Paddy, the rear gunner, who was a sort of human jackdaw, presented me with a hammer, a chisel and several keys which he'd found and assumed would be of use.'

Dudley was quickly drafted in to help run the prisoners' affairs. Although a flying officer, he was senior British officer (SBO) and was moved into a room with the senior USAAF officers.

The guards were all Romanian, mostly middle-aged with shoddy uniforms and patched boots. Dudley developed a good relationship with the camp commandant, a 'short, obese colonel', whose knowledge of French was as limited as Dudley's, but as none of the POWs spoke Romanian he was appointed interpreter.

He was also made camp supply officer, with his main duty issuing toilet paper. This highlighted differences between the British and the Americans who, he remembered, 'could not appreciate shortages – the majority of them had only recently left the US – and went through toilet rolls at a rate of knots'. As a result, he had to introduce rationing.[13]

Of the 205 Group POWs, F/O Douglas Calvert of 40 Squadron arrived at Camp No. 13 after being shot down on 7 May. His diary tells of grim life in captivity. 'I understand that conditions in the NCO barracks are very bad indeed – food very, very poor, about 100 men to a dormitory, which is very unhygienic. I hear that one or two of the men are cracking up a bit through nerves – there are no air-raid shelters – one or more cases of food poisoning thought due to bad bread containing hair, flies and often maggots.'

Some NCOs were from 614 Squadron. F/S Norman Foster RNZAF, a member of Dudley Egles' crew, recalled the conditions. 'Our ration of rye bread and some mysterious substance called mămăligă, a sort of pease porridge made from maize, was supplemented occasionally by a little meat and melted cheese. We kept as fit as we could by walking about the compound and playing soccer and basketball.'[14]

By late November 1944 the Russians were advancing towards Budapest, but for Syd Barlow and the survivors from JP323 it would be too late. They were taken from Budapest on the last train to leave before the bridges over the Danube were blown. Their destination was an interrogation camp, or 'Dulag Luft', at Wetzlar near Frankfurt. A Dulag Luft, a contraction of *Du*rchgangs*lag*er der *Luft*waffe, was a transit camp for airmen. There, they were issued with underclothes, razors, cigarettes and chewing gum, provided by the International Red Cross.

From Dulag Luft, where Syd bitterly recalled the Germans took his warm Irvin flying jacket, they were sent to Stalag Luft VII in Bankau, Silesia.

This camp for RAF NCOs had eight large huts, four on each side of a parade ground, Syd recalled. The latter was used for football and recreation. Each hut had rooms either side of a central corridor, each averaging about twelve men to a room with about 20 rooms to each hut.[15]

Geoff Curtis, meanwhile, was taken by train to Stalag XVII-B at Krems Gneixendorf, north of Vienna. His captors were confused as to why the three from JP223 had been in Hungary. Under interrogation, he was asked about being a Pathfinder and was accused of being a spy. Having been given a soldier's uniform after delousing, he claimed to be a soldier. Just after Christmas he was moved to Stalag XVII-A at Kaisersteinbruch – where soldiers were held.[16]

Amid the grim POW life, there were episodes of light. With the shortage of toilet paper, F/O Calvert recalled there were pictures of the Romanian King, Queen and Prime Minister, garnished with paper flags in the national colours of blue, yellow and red. 'Just the thing,' he remembered 'very soft and gave a certain psychological satisfaction'.[17]

Syd Barlow befriended their guard who spoke surprisingly good English. Syd remembered, 'He was a short, elfin sort of a man, with a pleasant, cheerful disposition. His name was Karl Grumm and he could speak English better than I could, and with a Lancashire accent!' He learnt it working at the Midland Hotel in Manchester before the war.

In meeting Karl, Syd and his fellow prisoners struck lucky. 'He was a marvellous man who would do anything that he could for "his boys".' The friendly guard dealt in the local black market and promised to get loaves for things like fountain pens and wristwatches.

A senior British POW, W/O Richard A Greene of the RCAF, was the so-called 'Man of Confidence' (MOC – a prisoner selected to liaise with the camp authorities) at the camp from June to December 1944. He recalled how, despite threats to guards for dealing with prisoners, barter got results.

The prisoners wanted to build a theatre, and work was eventually allowed but was slow because of materials which were carefully checked. Tools were almost impossible to obtain 'until cigarettes brought some marvellous changes'.[18]

There were many chances for POWs to get one over on their captors. One of these was the POWs pantomime. Syd recalled, 'The production was a hotchpotch of *Cinderella*, *Puss in Boots* and *Jack and the Beanstalk*.' The commandant said he must approve the script and there must be no derogatory remarks about Hitler or the German people.

The script was approved with the exception of the incidental music, which was the overture to *Merrie England*. The commandant said, 'England was no more merry than Germany, and that they should choose Wagner, Schubert or Strauss.' The MOC protested, saying the music sheets came from the Red Cross; they'd rehearsed it with this music, and 'we were damn well going to have it in our pantomime!'

An impasse ensued. The next day, a solution was proposed. The MOC said that the music was written by a German, and showed the commandant the music sheets, which said 'Music by E German'. This gave the opportunity for the commandant to climb down, and he grudgingly agreed to the music. They kept *Merrie England* for their panto and a small, but important, victory was notched up by the prisoners.

But with light there must come shade, and many times the POWs saw or experienced harrowing things which would mark them for the rest of their lives.

Life for the prisoners was hard but tolerable. But, one day, it became clear that brutality lurked close-by. On 27 December, Syd recalled an air-raid siren sounding but, because there were no shelters, the men were confined to barracks waiting for the all-clear.

Three of the eight huts had no toilets, with the only ablutions on the other side of the parade ground. The men heard the all-clear three hours after the warning and this, Syd recalled, resulted in 'a small cavalry charge of men' running across the parade ground to relieve themselves. Syd was almost in the vanguard of the race when they heard on the wind a faint 'Halt'. 'Suddenly one of the leaders threw his arms in the air and I heard a shot... he was about twelve feet from me and, when I looked, it was obvious that he had been killed instantly. His name was Lees Stephenson, a Canadian. I don't suppose he ever felt what had hit him, but it was tragic that he had been killed with no excuse other than he wanted to pass water.'

While sometimes seeing the worst of human behaviour, the POWs often also saw the very best. After escaping execution in a Hungarian village, Syd Barlow and his crew were taken to Budapest to be held in a local jail. In their cell they were startled to hear, 'Psst! Comrade!' Tracing the voice to a ventilation grille, they saw a face in the neighbouring cell. He then put his arm through a slit and handed them sandwiches of white bread with chicken, two small jam rolls covered with chocolate, and an apple. 'Thank you, Comrade, thank you!' the starving airmen chorused.

The face across the way said, 'Morgen Kaput [tomorrow dead],' pointing his finger at his temple and saying 'Bang'. Syd was eating the chicken sandwich but, as soon as he realised it was probably his last meal, he couldn't swallow it. He recalled that none of them could sleep that night through hearing 'the poor bastard' pacing his cell. There were other sobering experiences, too.

It was the strange privilege of some 614 POWs to see the damage their own raids had caused, or to find themselves running for shelter as aircraft, including those from 614, attacked from above.

On their journey to Austria, Syd Barlow and his crew went through a town called Hegyeshalom. They'd marked targets there on the night of 20 September. He clearly recalled it as being 'one of our best efforts'. The raid had happened two months before, but they still were awed by the scenes of destruction.

Everywhere there were trains and wagons lying on their sides and only one track operated through the town. 'It was the only time we had come face to face with the terrible carnage we'd created.' Ron, Syd remembered, was elated at first, 'but, like us, became sombre as the realisation of the terrible consequences of war unfolded, and we soon became very quiet.'[19]

The POWs in Romania often saw not just the effects of the

bombing, but were blamed by locals for it. In July, Douglas Calvert wrote of having 'numerous foul looks and fist-shaking from civilians', after one raid. The next month saw heavier raids on the oil refinery at Ploiesti. On 9 August, Calvert was disturbed to see RAF bombers come under attack. 'One poor chap, in the searchlights for five minutes, seemed totally lost, everything let loose at him, and he was attacked by fighter. He was believed to have gone down to the south of the city.'

The POWs in Bucharest had little protection from the falling bombs. Dudley Egles recalled their 'shelter' – a semi-basement dining room – 'which was better than nothing'.

One night, he said, the camp seemed to be 'surrounded by red flares – target-markers'. He knew exactly what they were and that they were dropped by his squadron. The camp was quite near to the marshalling yards which appeared to be the target. He observed, 'As reds usually signified the place to bomb, it was rather worrying. A few bombs did drop fairly close to us, but thankfully didn't hit the camp.'

Despite taking on a range of camp duties, Dudley Egles found there was little to distract from prison life's tedium.

'Overcoming boredom was our greatest problem. There were a few books, but I managed to keep occupied with my duties as SBO and interpreter, giving navigation lessons to the Yanks, rehearsing for the camp concert and taking my shift at tunnelling.'

Escape attempts not only alleviated boredom, but kept spirits alive. One of his many roles was vice-chairman of the escape committee. One of their projects was a tunnel, and he recalled, 'Some chaps found a door behind the stage in the assembly hall. This door was nailed shut and swathed in barbed wire, which was an invitation to see where it led.' The door was prised open and found to lead to a small semi-basement room. 'It was decided,' Dudley said, 'that this would be a good place to start a tunnel.'

The tunnelling went well until they came across a wall

about two feet thick, reinforced with metal rods. Using the hammer and chisel given to him by Paddy Beevor, a start was made, but when these were struck the whole building seemed to reverberate. The tunnellers disguised the noise by a carefully-timed loud clattering of pans in the kitchen after meals, and once they dug through the wall, more progress was made.

Once finished, they 'got a few bods out', Dudley recalled. Once, he managed to leave himself, but was free for only a few hours. He was hampered by his handlebar moustache which looked rather out of place in Romania, where few men of his age (23) sported facial hair. He sadly recalled that all the escapes were only ever temporary and none managed to get completely away.[20]

Another way of undermining prison authorities was acquiring a radio. Once a set was assembled, updates from the outside world on the progress of the war could be received. On the pretext of wanting to invite the camp commandant for tea, Dudley managed to persuade his captors to let him out to do some shopping to buy a teapot. He hoped that while outside he'd be able to obtain some parts for a radio. This he did, and by August they finally listened to news from home, and some of that news was good.[21]

On 9 August, the day of a very costly raid for 614 over Romania, Douglas Calvert remembered, 'We heard that Romania and Bulgaria were asking for peace terms, so what would become of us?' Only twelve days later he heard how the picture had improved. 'Heard the news that the Russian offensive on the southern sector had started in earnest. This was really good news for us, maybe it wouldn't be long now.'[22]

The 23rd of August was, Douglas wrote, 'a fine day'. Late in the day he heard on the radio that Romania had accepted unconditional surrender terms.

He had doubts about their chances. Later they saw more guards arrive, machine-guns erected, roads barricaded, although he wrote, 'evidently these precautions are to defend us'. There then followed a few days of tension, confusion but sometimes elation as the old order crumbled. More Romanian soldiers arrived amid fears that the retreating Germans might come. The Commandant now hailed the prisoners as allies, and said they should unite against the Germans. An advance party of Russian soldiers appeared, causing celebration. Despite warnings not to, many prisoners left the camp to visit nearby Bucharest.[23]

By 24 August their situation was becoming more dangerous, however. German aircraft bombed the city and plans were made to evacuate the camp, until it was realized that German troops had actually arrived nearby.

A few days later, a plan was being put into place which would get them all out of there. It was bold and highly risky, and even the planning of the operation itself was fraught with danger.[24]

Dudley Egles recalled the daring scheme which would see an American POW being flown to an Allied base in Italy by a Romanian pilot flying a German Me109.

Formulated in the days immediately following the armistice, the Senior Allied Officer at the POW camp, US Colonel James A Gunn, approached the Romanian Foreign Minister, Georgescu, asking for help in evacuating the ex-POWs who'd been released.

Georgescu introduced Gunn to a Romanian Air Force officer, Captain Carl Cantacuzino, who was persuaded to fly Gunn to Italy. They flew a stripped down Me109 with American Stars and Stripes painted on each side and carrying their evacuation plans for the Allied aircrews. They flew without oxygen, with Gunn lying prone in the radio compartment. Once at Bari, Colonel Gunn was able to organise B-17 bombers with a fighter escort to fly to Bucharest to pick up well over a thousand American and British POWs.[25]

The plan was quickly communicated to the camp – now

with its own wireless room connecting it to Allied Command in Italy. At a meeting in the CO's room, plans for a mass evacuation were announced to cheers. It would begin the next day, 31 August.[26] They left the camp early and got to nearby Popesti Airfield, taking off just as the Russians were arriving in Bucharest.

On arrival back in Italy, the Americans received a hero's welcome. The Brits, Egles sadly recalled, had no such return. 'We arrived back at Bari to a great reception of brass band, film cameras etc., but not for the non-Americans. No-one seemed to want to know us.' Unable to find help from the British authorities to get back to Amendola, he eventually loaned an American jeep and drove to 614's home himself.[27]

The squadron diary told how some of the former prisoners returned to the unit. 'News was received during the month that crews previously reported missing over Romanian targets had been evacuated to Italy on the signing of the Russia-Roumania [*sic*] Armistice.'

It added, 'Some of them paid a visit to the squadron where their stories of their experiences were received with great interest. They had been reasonably well treated, their biggest complaint being the quality food (soup, bread and ersatz coffee) and the lice, bugs and fleas which abounded in their prison camp. It was the first time that the squadron had had any news of some of the returned personnel and everyone was delighted with the good news. Most of them are now on their way to the UK.'[28] Dudley Egles was one who'd be going home.

As the Romanian POWs were flying back to freedom, those shot down over Hungary might well have cursed their fate. Hungary, too, would soon attempt to sign an armistice with Russia, but most of 614's POWs went to Germany before this happened.

Their captivity would go on, and the final months of their ordeal would be spent enduring a forced march across a frozen central Europe. January and February 1945 were among the coldest winter months of the 20th century in Europe, with blizzards and temperatures as low as minus 25c. The POWs were ill-prepared for the evacuation, having suffered years of poor rations and wearing thin, threadbare clothing.

The 614 prisoners still in German prison camps would face this awful march, and their harrowing experiences would live with them until the day they died.

Among the first of the 614 prisoners forced to march west was Syd Barlow and his crew. His camp, Stalag Luft VII, was in Bankau. By January 1945, Soviet forces were fast approaching and the camp authorities were told their POWs must march westwards.

The British POWs' leader, P/O Peter Thomson, RAAF, described the departure and what the men could expect on the march.

'On January 17th, 1945, at approximately 11am, we received notice of one hour in which to pack our kit and be ready to leave the Camp by marching. At the same time we were informed by Oberfeldwebel Frank [the German camp commandant] that for every one man who fell out of the column on the march, five men would be shot. This order was never given in writing.'[29]

Fortunately, Syd recalled, the start was postponed until 3.30am on 19 January, giving him time to make some preparations. He built a sleigh from an Argentinian Red Cross box and bed planks, with a rope made from string he'd acquired.[30]

Thomson told how each man was issued with two-and-a-half days'worth of rations before leaving. No transport was supplied for any sick and there was little medical equipment.

The 1,550 men left Bankau, and marched 18 miles to Winterveldt. Thomson recalled, 'This was done under extremely trying weather conditions and severe cold. The only accommodation at Winterveldt was small barns.'[31]

The two following days were hellish. He wrote, '20 January: marched from Winterveldt to Karlsruhe, arriving at 10am. Set off at 5am and marched 8 miles. At Karlsruhe, housed in abandoned brick factory. Here, for the first time, we were provided with two field kitchens with which to cook for 1,550 men.'

They were given a horse and cart to transport the sick. The same night, despite protests, they were ordered to move and left Karlsruhe and marched to Schönfeld, arriving at 9am on 21 January, covering 26 miles. The temperature in the night dropped to minus 33c. The medical officer's wagon was filled after the first three miles and, from then, men were being picked up at the roadside in a collapsed and frozen state. After crossing the River Oder and covering 21 miles, they were told that they could rest for two days.

'21 January: at Schönfeld, accommodated in cowsheds and barns. A room was provided for the sick at Lossen.'

There would be no respite. The exhausted men were roused in the middle of the night from what sleep they could snatch on the cold floor.

'22 January: at 3am, orders were given to prepare to march off. It was dark and there was some delay in getting the men out from their sleeping quarters because they could not find their baggage.'

To hurry the prisoners, the guards marched into the quarters and fired their weapons. The column was marching again by 5am. 'Twenty-three men were lost. They may have been left behind asleep, or they may have escaped. We marched to Jenkwitz, a distance of 21 miles, and were housed at a farm in barns.'

After this brutal awakening and a 20-mile march, they were at least given more food – 200lbs of fat, 46 tins of meat, barley, peas, and three-quarters of a pig. The next day, they left Jenkwitz at 6am and marched 13 miles to Wanzen.[32]

Syd Barlow wrote of those brutal days, trudging through

ice and snow. 'Conditions were horrendous, it had been snowing on and off for over a month, with snow up to six feet deep on the roads. Blizzards came almost every day, and you would find yourself staring and unable to focus your eyes. Every step was mechanical, but a victory if you could keep going.'[33]

On 24 January they were rested for the day at Wanzen, and the men slept in barns. After marching 19 miles to Keidersdorf on 26 January, they were given 600 loaves of bread to last for two days. The next day they left Keidersdorf and marched twelve miles to Pfaffendorf, arriving at night. They left Pfaffendorff for Standorf at 05.00 on 28 January and marched another 13 miles.[34]

Beyond these bald statistics lay stories of unimaginable hardship. The gnawing hunger, the savage cold, the brutality of the guards. Although exhausted, sleep was difficult. Syd recalled, 'When we sank down to rest in the barns, it was difficult to sleep through the noise of coughing by nearly all the prisoners. There were several cases of dysentery and almost everyone had frostbite.' They would then be roused by the guards to move again, which invariably 'was in the small hours'.

Death stalked them. 'A number fell asleep on the march, froze, and never woke again.' But there was also extraordinary selflessness and sacrifice. 'I remember the exceptional gallantry of Padre Revd Jack Collins, a tall giant of a man who carried one prisoner on his back for almost an entire days' march.'

As they staggered on, something miraculous and powerful happened to Syd – proof of the capacity for kindness amid this utter misery.

He admitted the episode was difficult to talk about, even after some decades. 'It sounds far-fetched, but it's perfectly true.' It happened when the march was some 20 miles from Cottbus.

The prisoners were halted, so stamped their feet and blew into their hands to keep warm, waiting to march on. There was a small farm on the right-hand side of the road and the farmer's wife was gazing at the throng from inside a wicker fence. 'She was a large lady,' Syd remembered – of similar proportions to his own mother. On impulse, he approached her and used his basic German to ask, 'Haben sie kein trinkwasser, bitte?' [Have you some water to drink, please?]

'She looked at me for what seemed ages, and a bewildering change of expression crossed and recrossed her face, until finally that face crumpled and she said, "Komm!" She led the way into the kitchen where it was lovely and warm and set me down at the kitchen table. She gave me what must have been a litre of still warm milk which she poured from one of those wide-mouthed ewers with blue and white bands on the sides. She also gave me a thick slice of white bread, buttered and covered with honey. I'll never forget the taste of it.'

Syd admitted 'he was overcome with her kindness'. He soon understood the reason why she decided to bring the disheveled prisoner into her home. 'She made me understand that she had a son on the Eastern Front but hadn't heard from him since 20 November.'

After he'd eaten, the woman pointed to a photograph on a shelf in the corner. Syd recalled, 'I went cold. The photograph could have been me except for the German uniform.' It really was uncanny, he thought. 'After this encounter,' he said, 'I prayed for some time that this lady's son might be returned to her.'

In that brief encounter their lives, so utterly separate, were joined and very briefly they gave each other comfort in the most unlikely of settings.[35]

And so, the ordeal continued. P/O Thomson chronicled the grim miles, early starts in the freezing dark, and the meagre rations. The next stage of their journey would leave them nearly broken. They marched the 13 miles to Peterwitz in darkness through a blizzard. They arrived utterly exhausted.[36]

There the prospect of transport was dangled. But they still had to march the seven miles to Prausnitz, then five miles from Prausnitz to Goldberg where they were squeezed into cattle trucks, 55 men at a time. They were on the train until the morning of 8 February.[37]

Their marching was now at an end, but Syd remembered this as equally harrowing. 'We had to sleep six or eight at a time in a sitting position, along one side of the truck, in a rota. The coughing was constant; we had no water for two days, and were only let out twice to relieve ourselves. A can was passed to urinate in and was emptied through cracks in the floor.'[38]

Peter Thomson summed up the effects of the ordeal, 'The men suffered from an extreme degree of malnutrition, and an outbreak of dysentery. There were numerous cases of frostbite and other ailments. They were quite unfit for any further movement. Food and better conditions were urgently required.'[39]

This was their state as they arrived at Luckenwalde on 8 February. They had to walk the final few miles to their new home, Stalag III-A. This was where they finished their captivity and where they were liberated by the Soviet Army on 22 April.

Meanwhile, at Stalag XVII-A, where Geoff Curtis was, their long march wouldn't begin until April. Research by the museum of Kaisersteinbruch, where Stalag XVII-A was sited, sheds light on the start of the move. 'When the Red Army approached Stalag XVII-A, like most of the camps in Wehrkreis [Military District] XVII, it was ordered to evacuate to the west.' Notifications were issued from the Swiss Foreign Office (as part of the neutral nation's protective power function) to the US State Department on 21 April 1945, which told of a 'transfer' of the camp to Braunau.[40]

Geoff recalled French, British, Czechs, Americans, Italians, Poles and Russians poured out. It was, he remembered, 'a Tower of Babel – under heavy armed guard'. Food for Geoff and the prisoners of Stalag XVII-A was as scarce as it had been for Syd Barlow. Prisoners were given meagre rations and after they'd run out, they had to forage for themselves. Geoff recalled that one enterprising New Zealander fashioned a hook on the end of a stick which he kept under his coat. When he saw an unsuspecting chicken wander across his path, he'd whip it out and snatch the unfortunate bird, break its neck and hide it under his coat to cook later when out of the guards' sight.

They became ever-more desperate for food. A dead horse in their path was eagerly cut up. Geoff remembered finding a sodden loaf of bread under a barn which he and others wolfed down. They marched westwards, only being allowed to rest when they were exhausted. Some of the Russians had no shoes and had tied sacking around their feet.[41]

On they went until they reached Braunau in Austria. Jean-Marie Marcel Meyssignac, a Catholic priest and Camp Chaplain in Kaisersteinbruch, recalled how they were guarded by 'Volkssturm' – or 'Home Guards'.

He said some prisoners lived the remaining days of their captivity in self-built huts, but salvation was at hand, he recalled. On the evening of 3 May, it was rumoured that the Americans were approaching. The next day they appeared, ending the captivity of Stalag XVII-A's prisoners.[42]

F/S Patriarca would also go through the hellish experience of a long march. At some time in the autumn of 1944, he was taken to Stalag Luft III at Sagan. Best-known for an escape plot in 1943 (depicted in the film *The Wooden Horse*) and the Great Escape of March 1944, Ron's family received confirmation that he was there on 1 January 1945. He was soon moved to Stalag Luft IV at Gross Tychow, but on 6 February Ron and other POWs were forced to march around

19 miles south-west to Stettin. He described prisoners' shoes falling apart as they walked in bitter cold on this harrowing three-week journey. With little food or water, the starving men were thrown occasional pieces of stale bread. Many perished due to sickness, hunger and exhaustion and he saw men dying at the roadside. Others were murdered by their guards. This march is known as the 'Black March'.

Sometime between the end of February and the end of March, Ron was forced to march again – this time for six weeks towards Hanover where he was liberated on 11 May.

Traumatised by these events, these spare details were all that Ron told his family, yet those words and the accounts of Syd Barlow and Geoff Curtis give an insight into the horror these men experienced on those long marches.

CHAPTER 9

Italy, the road to victory (1944–1945)

ALTHOUGH RAIDS ON oil refineries continued after Romania's armistice, the balance of 205 Group attacks shifted to enemy transport networks. That mostly meant attacks on bridges and harbours and railway marshalling yards.

Another focus of the campaign was the River Danube. This vast waterway, flowing from Germany through the Balkans to the Black Sea, was the ideal conduit for supplies through Axis countries.

It was estimated that the Germans needed to move 12,000 tons of supplies eastwards every day to support the fighting against Russia. As the war progressed, and Allied bombers attacked the railway system, the River Danube became Germany's lifeline and was crucial to their war effort.[1]

One type of barge used on the Rhine could transport the equivalent of 100 ten-ton rail wagons and, by March 1944, the majority of all oil from Romania was transferred to the Danube and traffic on the river increased by 200 per cent. Even a temporary stoppage would have a huge impact on the German war effort, and the Allies planned to interrupt this by laying mines.

The result was a campaign which was one of the least known, yet most effective of those waged by the RAF in the war. Crews of 205 Group ran the gauntlet of night-fighters, flak and balloon barrages to sow mines at low level on moonlit nights. On these 'gardening' operations between April and October 1944, they drastically cut the movement of Axis river shipping and, at times, halted it altogether.[2]

At first the missions were flown only under a full moon because the bombers had to fly at no more than 200 feet, and sometimes as low as 40 feet in order to accurately place their mines. To allow them to be sown at any time, the Pathfinders of 614 were used to illuminate the target.[3] Missions on 1–2 July, the biggest up until then, saw 16 Liberators and 53 Wellingtons lay 192 mines, and 60 more on the following night, with five 614 Halifaxes acting as route markers for the mine-laying bombers.[4]

On 10 September, a solitary 614 Halifax joined 17 Wellingtons of No. 231 Wing, and eleven Liberators of Nos 240 and 2 SAAF Wings. The 614 Halifax, JP279, dropped yellow TIs and flares, allowing the 25 mine-laying aircraft to 'sow' 82 mines in four 'beds' between Esztergom and Szap. The risk of these operations was shown 20 minutes later when a yellow flash, followed by two small explosions and a fire, were observed. These were believed to be from a downed aircraft. The 614 Halifax returned safely,[5] but one Liberator of No. 34 Squadron SAAF was missing.

By August, reports showed the results of the operations, cutting oil traffic 70 per cent with at least 150 vessels sunk.[6]

Supplies to the German armies in south-east Europe were drastically reduced, and later virtually ceased. Salvaging wrecked barges from the shallow river also created huge disruption. The Germans admitted their struggle 'to clear the Danube of mines with the means we have at hand and the position regarding shipping is badly affected in consequence'.[7]

The last 'gardening' operation was flown on the night of 4 October 1944, when Liberators and Wellingtons laid 58 mines, bringing the grand total laid to 1,382 in 18 attacks.[8]

That autumn, 614 squadron's accommodation improved. From having to sleep under canvas since reformation in March, following the move to Amendola the airmen attempted to make their living and recreation quarters a bit more resilient in preparation for the expected bad weather.

The early part of September saw huts for the three messes, Navigation, Briefing and Headquarters all finished. The diary admitted, 'The huts are very rough, but will at least be dry in the winter. The airmen's canteen and recreation hut was also finished.' Built entirely by squadron labour it would be 'greatly appreciated in the bad weather to come'.

The onus was on the squadron to provide for itself. 'The airmen have used their brains and initiative in constructing for themselves a shelter that will at least keep out the wind and rain,' the diary told. The results were practical but not stylish. 'The variety of materials used has to be seen to be believed – scrap wood, corrugated iron waste, flare box lids etc.,' as building supplies, huts and tents were very scarce. 'The whole camp looks like a Boom Town in the best Hollywood tradition.'

To make plain the need for more solid shelter, the second half of September brought fierce weather with cold, rain and gale-force winds. There was one night which 'caught the camp unawares' and the following morning's casualties included the sergeants' Mess, A, B, and C Flights, Photographic Section, Telephone Exchange, Post Office and six domestic tents. 'As a sample of what is to come,' the diary noted grimly, 'the future is not very bright'.

By their own efforts, recreation for the airmen was improving. In September, the cinema was holding nightly performances, and on 4 September the hundredth show was given. Morale was 'generally high', with one big factor the recent improvement in quickly posting tour-expired airmen home.

Morale would have been further boosted by the kick-off of the No. 205 Group football league on 20 September, with a game against No. 37 Squadron, which 614 won 3–1. The forthcoming matches were being looked forward to 'with great interest' and ten friendly matches were played during the month. Plans were afoot to set up a league for rugby, too. There was also the award of a DFM to F/Sgt FW Beare for 'gallantry in flying operations'.

Meanwhile, in early October, a momentous shift began when the squadron began to convert to a new aircraft – the American Consolidated Liberator. In many ways it was a drastic change. Powered by four Pratt & Whitney Twin Wasp fourteen-cylinder air-cooled radial engines, each giving 1,200hp, its long, thin Davis wing, twin rudders, tricycle landing gear and box-like fuselage, made it easily recognisable.[9] While in many ways a more capable aircraft than the Halifax, with greater range and better defensive armament, many were sad to see the back of the British bomber. Not least Tom Scotland, who bade an emotional farewell to the Halifax he had named 'Kathleen', which he flew to Algiers to return with a Liberator.

'She looked so proud but so alone,' sitting on the tarmac at Maison Blanche, and he recalled crew-member Tony Wheeler christening her by swung a bottle of sparkling wine while against her side. 'Farewell!' he said to the aircraft that had got them home so many times.[10]

Nevertheless, Scotland was impressed with the Liberator. Much about the new bomber exuded power, comfort and plenty – exemplifying what the industrial might of America brought to the war. Tom was astonished by the comfort, saying the pilot's seat felt 'like a lounge chair'.[11]

It may have been spacious compared to the Halifax, but the Liberator was not without its foibles. The crew usually entered via the bomb-bay, simply ducking to clear the lower edges of the opened bomb-bay doors (designed like a roll-top desk cover). Hoisting themselves onto a 'catwalk' which extended the length of the bomb-bay, they walked along it to access the the forward section of the bomber. Those stationed in the nose, like the bomb-aimer and navigator, could enter the aeroplane by crouching under the nose and squeezing past the nose wheel tyre and strut via the nose wheel door opening.[12]

Nevertheless, its rate and scale of production was equally impressive, indeed staggering. The largest factory to produce the B24 was Ford's specially-built Willow Run plant in Detroit. The mile-long factory produced a third of the 19,000 Liberators built between 1940 and 1945.[13]

Another member of Tom's crew had misgivings about the new bomber. Before moving to Amendola, the squadron had been based alongside many USAAF units at Celone, which flew the Liberator. To Navigator Bill Scott, the type seemed accident-prone. He recalled, 'On returning from an operation, a formation of American Liberators flew over the airfield and started to "peel-off" before landing. Unfortunately, as the first aircraft left the formation, the aircraft adjacent to it lost an engine and slipped into it, one crashed into the runway on the one side, the other crashing on the opposite side, with the loss of about 18 lives.'

He remembered that Tom Scotland spent about two weeks learning to fly Liberators, then the whole crew took the next four weeks doing familiarisation flights using H2X (the American equivalent of their H2S blind-bombing radar).

He had other gripes with the American type. On the Liberator Mk VIII, the Navigator was sat immediately behind the Pilot in line with the engines. 'The chart table was less than half the size of that in the Halifax, and the least vibration had the table bouncing and it was impossible to keep anything on the table.' As his pen leapt across the plotting table he would plaintively ask Scotland, 'Navigator to Pilot, please try to synchronize the engines.'

He remembered their first Liberator mission on 12 October 1944, attacking Bronzolo's marshalling yard in northern Italy. The yard was a vital hub controlling access to the Brenner

Pass – the main thoroughfare between Germany and Italy. He recalled they dropped flares 'blind', using the American H2X to locate the target.[14]

Halifaxes and Liberators operated together. Tom Scotland recalled how strange it was to watch the seven British-made bombers take off as he waited in the new Liberator. As they neared the target, shells exploded just off their starboard wing. The shrapnel fell on them, pattering like rain.

As they flew in over Bronzolo, Tom recalled peering down into the darkness of a deep valley far below where the unseen target lay. They dropped flares on the target, and from the darkness came a view of 'railway sheds, lines of railway trucks, row upon row of railway lines'. More Halifaxes from 614 went in to drop their red and yellow target indicators, and the target 'became bathed in cascading colour' for the bombers following them.

Meanwhile, the squadron's first XI football team were on a winning streak, with victories over 178 Squadron, 2933 Squadron RAF Regt, and 240 Wing, only losing by the odd goal to No. 70 Squadron. The squadron's rugby team began serious training, and were to play their first match early the following month.[15]

Bridges, too, became a focus and the squadron attacked many in November, with the first mission of the month against the Latisana railway bridge in Italy. Then, good illumination helped bombers attacking the bridge at low level score a direct hit. More bridges were attacked in the Balkans, with targets at Lješ in Albania and Visegrád in central Yugoslavia. There were further raids too against Latisana and Tagliamento/Casarano railway bridge in Italy.

Over Christmas there was little rest, with many Italian targets being hit. On Boxing Day, four Liberators and four Halifaxes attacked Casarano bridge, and the next day Piave/Susegana railway bridge. Flares dropped gave 'excellent illumination' with the TIs concentrated in the centre of the bridge. On 28 December, Casarano bridge was hit again.[16]

This high tempo of attacks eventually told on the German supply chain. That winter, across the whole of northern Italy, German records showed 1,020 cuts to transport links, of which 754 were made by aircraft attacks, 154 by partisans and 112 by accidents. Of the breaks to transport links, 189 were of bridges. One agent's report stated that road traffic, in the first week after an attack along the east bank of Lake Garda, had increased to 750 motor vehicles southbound and 400 northbound.[17]

Alongside the missions against transportation and other targets in December, Christmas 1944 saw 614 bring good cheer to locals. On 23 December the squadron entertained children from a local orphanage with an afternoon dress rehearsal of the squadron's pantomime. It seems, as is often the case with panto, that some of it went over the youngsters' heads, and the diary records how 'their reactions were very amusing – not understanding the language – but the slapstick parts were well appreciated.'

After the show they were taken to the airmen's canteen, where a chain of them carried the children across the mud and they were given tea, sandwiches, cakes and sweets. The party finished with the children singing 'Silent Night' for their hosts, and the diary writer saw how the airmen who had families of their own 'probably enjoyed the party more than the children themselves,' and no doubt were mindful of their own families at home.

On Christmas Eve, the squadron's choir went round the camp, singing carols, led by the Padre collecting for the POW fund, and 'provided a real Christmas touch'. A double-feature programme was shown at the cinema and attracted a large crowd. Christmas Eve finished with a midnight service held in the squadron's church.

Christmas Day was cold, wet and very muddy. There were no operations scheduled, but some crews would be told differently.[18]

Tom Scotland recalled the unexpected news which greeted them. With a greatcoat warming him against the biting cold, he walked towards one of the snow-covered seven-man tents which still housed his crew, despite their efforts to build wooden shelters.

Under one arm he carried a Christmas food parcel from home. He imagined the excitement of his crew when he opened it. Suddenly, a voice called 'Scottie!' He turned to see the adjutant calling him from across the snow. 'Operations are on; your crew is needed at briefing straight away!'

Three hastily assembled crews sat in an icy tent for the Christmas morning briefing. They were to fly to the mountains of southern Germany and make an assault on Adolf Hitler's secret bunker at Berchtesgaden. Intelligence revealed that Hitler was likely to be at his Alpine hideout. 'They'd deliver a "Christmas parcel" from 614,' as Tom remembered it. Part of the route had to be flown at low level to avoid enemy radar. With snowstorms sweeping across Europe, the plan was very risky.

Gripped by apprehension, they left the tent and ate a silent breakfast as they pondered the long, dangerous flight ahead of them. In the truck they bumped along the rutted, icy track, arriving at the airfield where they were met by a scene of furious activity.

Groundcrew with brooms cleared away ice and snow, sweeping the runway, the taxiway, and clambering over the great bombers to brush snow from their wings.

Scotland's crew clambered aboard while he walked around the Liberator, doing pre-flight checks. When he started the four engines, the throbbing song of their power helped to settle him.

In the cockpit, he looked out for the green light signalling them to taxi out for take-off. Then something made his stomach tighten and a shout of disbelief came. 'Bloody hell!' cried Ed.

Rolling down upon them, a black cloud enveloped their Liberator, rocking and juddering the giant bird and blocking the windscreen with snow. Time swirled by as they sat waiting, watching the circling hand of the ticking clock. Eventually the storm passed, and Tom opened a side window of the cockpit to gaze out on a scene of complete whiteness. Nothing but whirling propellers moved. The swept surfaces of their aircraft and the runway again lay under snow.

A snow-covered jeep approached and stopped in front of them. The driver stood up, raised his arms, and Scotland recognised the operations Wing Commander telling them to switch off engines.

The crew sat in silence, surrounded by thick snow. Tom clambered out and the groundcrew sergeant met him. 'Operations are cancelled, Sir,' he said in Scottish brogue. 'Are you sure, Sarge?' Scotland asked. 'The Wing Commander says so, sir,' he replied. 'Happy Christmas,' Tom said, with a smile. He called along the aircraft, 'Operations are cancelled fellas. Happy Christmas!' Ed collected his gear and muttered, 'A bloody silly idea anyway!'

They later found out that an informant in Germany had learnt of Hitler's changed Christmas plans. He would stay in Berlin. The Führer's Christmas parcel would need to be delivered another time. 'Come on blokes,' said Tom as they left the Liberator and their flying boots crunched on the snow. 'Let's go and enjoy our Aussie Christmas parcel.'[19]

With operations scrubbed, all could enjoy Christmas. Christmas dinner was served, according to custom by the Officers and senior NCOs.[20] Hugh Fleckney recalled they were each given two bottles of Tennent's Lager – a great treat as beer was something of a rarity.

Boxing Day brought fine weather and some operations took place. The squadron's show, 'Pantomime, Fun and Frolics', produced by the adjutant, was presented in the cinema to a packed house. The squadron's spiritual home in south Wales remembered 614 with a gift of £25 from the County of Glamorgan Territorial Association which 'was very much appreciated and suitably acknowledged'. On New Year's Eve the festive spirit returned, especially with the Scottish contingent celebrating Hogmanay.

December also saw 18 Pathfinder badges awarded to squadron aircrew and, the previous month, AOC 205 Group, Brigadier Durrant, also presented Pathfinder badges to squadron aircrew. Their award meant that 614 was finally fully recognised as a Pathfinder squadron. Referring to 614 Squadron, the Air Ministry record states, 'Employed in target marking role. Has been granted Pathfinder establishment and privileges similar to those applying to Pathfinder squadrons in No. 8 Group, Bomber Command.'[21] Then, other awards were announced, including DFCs for 'gallantry in flying operations' for Flt Lt DQ Watson, P/O BV Ross, P/O RD Francis and F/O SE McFall.

Some weeks earlier the squadron had begun producing its own newsreels to be shown in the camp cinema. The first would be shown at Christmas. The diary told, 'A good start has been made; already all the necessary equipment has been obtained, including cameras. Subjects included the parade where Pathfinder badges were awarded and a comedy piece showing one of the cooks chasing a turkey.'

New year saw more bad weather and the risk of accommodation being flooded. Happily, preparations had been made and a wash-out was averted. However, roads and tracks around the camp were still turned into a quagmire.[22]

The new year would see the squadron remain focused on hitting transport targets and, often, crews would be joined by aircrew under training from the UK. For a mission to Udine in northern Italy on 20 January, Tom Scotland's crew had a new member.

Tom recalled they approached the target at five minutes past midnight. 'Our new bomb-aimer, Ted Budd, whom Ed

Riley continued to train, lay in his position to guide me in over the target, ten minutes away. Ed sat gazing at the radar picture of the target area.' As they flew in, he guided the new bomb-aimer, 'OK, the target is 35 miles away, we are dead on course.' Exploding gunfire indicated the first 614 aircraft had reached the target. Bursting light showed their sky flares hanging in the sky on parachutes, and looking for all the world like 'a Christmas tree hung with bright candles'. The attack on Udine had begun. Then, through the intercom, came an excited voice, 'I can see where we are,' said the bomb-aimer.

Ted began to guide Tom in. 'Left…' Tom adjusted the giant bomber's position by tiny degrees at the bomb-aimer's word. He noted grimly that the guns had begun to get their range. 'Left,' came Budd's voice again. 'Steady.' 'Gunfire close astern!' the top gunner shouted. 'Not long now, gunner,' Tom called. He was anxious lest the gunfire put them off their run. 'Steady… steady…' droned the voice of the bomb-aimer. The moments ticked by. 'Steady… steady.' 'Bombs gone!' They flew on. 'Camera flash gone,' Budd called.

Scotland turned the plane out of the line of gunfire and circled around to see Budd's green and yellow TIs glowing where they had fallen. They looked good, Tom thought. 'Can you see them, Ed?' he called as he gazed past a faint reflection of light from the ground, outlining the nearly vertical wings of their turning aircraft. 'Yes, I see them,' replied the Geordie. 'Bloody good.'[23]

Although 614 Squadron's main job was to mark and destroy targets, the bellies of its aircraft were often filled with more than pyrotechnics and bombs.

Between October 1944 and February 1945, 614 dropped thousands of tons of supplies to partisans in Yugoslavia. They also carried troops and equipment when Britain intervened in the Greek civil war in late 1944 and after VE Day.

Tito's Communist resistance consistently out-witted German attempts to crush them, and much of their equipment was parachuted by bombers of 205 Group based across the Adriatic from Yugoslavia.

Tom Scotland gave a sense of these risky missions. Flying low, often in poor visibility, they had to find and mark the drop zone, release their containers, aiming for forest clearings or torches in the dark.

But, beyond the danger was the knowledge that there were people who were grateful for the risk they ran and looked up to them with appreciation.

He recalled, 'We flew under clouds to a tiny clearing on a tree-covered hillside. A thin wreath of smoke ascended from the ground and spiralled into the air. Our drop point. With bomb doors open and feeling very vulnerable, I flew low over the small area of cleared ground and kept myself ready for possible German gunfire. Ed [Riley, bomb-aimer] pressed the button for the containers to drop and they were counted out one by one.'

Then they saw a figure.

'We caught sight of someone standing beneath overhanging trees, leaning on a long stick and looking at us. We knew that shortly after our drop other partisans would come out of the bush and drag the containers into hiding.'

Another time, one of those figures on the ground waved in gratitude.[24]

No. 614 Squadron made its first air-drops on 29 October 1944, over Bileća in Bosnia. Each drop zone was code-named – this was 'Soapsud'. Three Liberators dropped 36 containers and the squadron diary recorded blithely that, 'No opposition encountered. Weather conditions good. All aircraft returned safely.' But behind that bland summary lay meticulous preparation and targeting, every bit as precise as for a bombing raid.

One end of the canister carried a parachute deployed by static line. The other end had padding to cushion the landing.[25] The cylindrical drums, 1.7m long and 0.4m in diameter, could carry weapons, ammunition, fuel, medical supplies and more. The parachutes were colour-coded so their contents could quickly be known without having to open them. They were red for ammunition, yellow for medical, light blue for food and water, white for general stores, and green for signals equipment.[26]

The Liberator's cavernous bomb-bay could easily carry this kind of cargo, and typically held twelve 'Type-J' canisters. F/O Les Sims, who flew these operations with 614, recalled canisters were housed on frames fitted to bomb racks where they'd sit on each side of the bomb-bay to be revealed when the 'roller desk' bomb doors opened.[27]

For this mission, the Liberators took off at 14.30 from Amendola, soaring over the Adriatic. Bile a was around 220 miles away with the operation taking three hours.

Sitting on the valley floor in mountainous country, it was reached by map reading and using the Gee navigation aid. Fires lit on the ground would mark the drop zone.

Nearing Bile a, the Liberators flew in, one behind the other, at precise intervals. Liberator KG949, flown by W/O Sadgrove, was first over the drop zone at 16.10, dropping its twelve containers in salvo. Flashing the agreed code letter 'E' three times, no reply came from below.

One minute later was Tom Scotland in Liberator KG945. They saw one bonfire at position code letter 'H', so dropped their twelve canisters 'in rapid train'. All parachutes were seen to open.

One minute later came W/O Stephenson and his crew in Liberator KG957. They saw two fires and dropped all twelve canisters and saw them being retrieved.[28]

Although the squadron performed other drops to Yugoslav partisans that winter, it was called on to take on perhaps an even more hazardous supply job further south.

Following the German withdrawal from Greece in November 1944, a civil war broke out between the Greek government and the Communist-dominated Greek People's Liberation Army (ELAS) who were supported by Tito's Communist partisans and the Soviet Union. Britain and America, however, backed the Greek government, and it was to their aid that 614 Squadron came in December 1944.

British troops in Athens needed reinforcement quickly and, on 10 December, arrangements began for 205 Group to transport 2,000 troops, equipment and ammunition to Athens. Unlike the Yugoslavia missions, they would have to land and unload at airfields. The aircrews, highly skilled Pathfinders as they were, would have to learn fast how to land heavily-laden aircraft under fire. Their lives would depend on it.

The first sortie came on 12 December when Liberator KH237 carried 20 troops to Kalamaki Airfield in Athens.[29] As fighting around Kalamaki intensified, 614 continued to fly in troops and supplies. Les Sims was the pilot of Liberator KG944 for the mission on 20 December 1944.

'We were trained to jettison bomb loads before landing,' he recalled. But this time they would have to land loaded with ammunition under the very worst of conditions. 'The rebels were happily mortaring the runway as we were coming in to land. We felt very vulnerable.'

So dangerous was the situation that they were told to fly straight in from the sea and land with no circuit. There was help at hand, however. 'As we came in, two Hurricanes took off and went to strafe positions in the hills overlooking the airfield.'

Les nursed the laden Liberator down onto the runway and prayed the tyres would withstand the heavy landing. They nervously scanned the hillsides to watch for mortar fire. 'Fortunately, there was only one burst, and that occurred about one-third down our landing run... a quick touch of brakes and we straddled the small crater the bomb had produced, with a little bumping and deep breathing to help us.' They had made it – just. Now they had to quickly unload the aircraft while hoping mortars wouldn't find them. They couldn't shut down the engines as to restart would take too long. But, as they unloaded, Les worried the engines would overheat.

The cargo was quickly off-loaded – now they had to get out. The rebels would have seen the bomber coming in and would have it in their sights. They hoped the RAF fighters had kept their heads down. 'All of the Liberator's 5,000hp was used to get to the end of the runway and take-off,' Les remembered. 'It screamed and blasted its way into the air.'[30]

They returned the following day, with crews including Tom Scotland's in Liberator KG949, 'Altair.' He, too, had a mission fraught with peril but also had an unexpected passenger. Group Capt Jarman, Operations Strategic Planner for the 205 Group, wanted to go to Athens to see the conditions for himself. They landed after RAF fighters silenced Communist guns bombarding the airfield. After rapidly unloading they quickly lifted off from the pock-marked runway. On the journey home Jarman confided that he was not pleased to be risking their bombers in such a dangerous enterprise.[31] It's not clear if, shocked by the risk, he called for the missions to cease. But, the next day, 614 flew its final supply run to Athens, having delivered 879 Army personnel and their equipment and over 24 tons of ammunition.

Although strategic bombing was 614's main role, alongside these transport missions, they also supported ground operations. From July to the run-up to the Allied invasion of southern France on 15 August, they worked with the French Resistance.

The change was appreciated by the crews. 'On the 14th, crews enjoyed a very interesting trip to the South of France – a diversionary attack on Valence Aerodrome, enabling French Patriots to destroy aircraft.'

The next raid, in support of the Resistance, was on 24 July, when seven aircraft illuminated Valence Airfield again. The squadron had another unexpected guest.[32]

The Group navigation officer, Arthur Spencer, attended the daily air staff meetings at Bari, alongside an Army officer who liaised with the French underground. Arthur remembered, 'The Army officer arrived one morning to inform us that the French resistance intended to attack an airfield in the Rhone Valley and would welcome a diversion to make the Germans keep their heads down while they went about their business.'

Arthur thought that this 'sounded a very interesting trip', so he contacted 614 and found out that there was one crew without a navigator and so he flew with them. A fortnight later he discovered how successful the attack had been. 'The Army officer announced that the Resistance had been well satisfied with the diversion provided and had destroyed 37 German aircraft on the ground, blown up the ammunition dump, and killed an unspecified number of Germans.'[33]

In late August, 614 flew raids in support of the 8th Army's attacks on the Gothic Line – the fortifications built by the Germans to block the Allied advance north through Italy. Between 27 and 31 August, 614's Halifaxes marked targets for bombers at Pesaro and the diary noted crews felt they were directly aiding the 8th Army with the attacks there.

The squadron also led raids on enemy airfields which, while appreciated, were costly. On 6 July, eight 614 Squadron Halifaxes flew to illuminate and bomb Feuersbrunn Airfield, near Vienna. The target was well marked but there was stiff opposition from 'many night-fighters' and the Group suffered heavy losses. Halifax JP287 failed to return, crashing at Innerhalpbach after being attacked by night-fighters. All the crew members were killed and they are buried at Klagenfurt war cemetery in Austria.

MASAF Chief General Twining congratulated 205 Group following the raid, but the diary also recognised 614's loss. 'We ourselves lost a flight commander and a very excellent crew. However, the results proved the effort paid. Nineteen enemy aircraft being destroyed on the ground.'

That crew were Pilot F/O Maurice George Harrison; Navigator F/O Glyn Davidson Rees; WOP F/S Walter Richard Vokes; F/Eng Sgt Norman Harold Hodson; Tail Gunner W/O Arthur William Pryor; Bomb-Aimers Flt Lt Alfred Sudden and Sgt Percy Charles Dolman.[34]

Another raid that autumn was on Szombathely Airfield in Hungary, and it took place on 21 October under a full moon – ideal for night-fighters. Tom Scotland recalled it was 'an open invitation for the Germans to attack and attack they did... playing havoc with our stream of bombers.' A Halifax was hit. 'It arched through the sky with flames marking its pathway to the ground.' It was from 614. 'At debriefing, we learnt that another of our key Pathfinders, Ian Bruce, was missing.'[35]

The crew of Halifax JP232 consisted of Pilot Flt Lt Ian Henderson Bruce; F/Eng Sgt Harold Dunn; Nav P/O James J Earl; Bdr P/O John McCormack; M/U/Gr W/O Peter McKee; WOP F/S Edmund Joseph Sharp.[36] The squadron mourned the loss of this 'very experienced and popular crew'. The only survivor was P/O Edward Charles Wearing, who was captured and became a POW, held at Stalag Luft III.[37]

In September 1944, German forces began to withdraw from Greece and 205 Group bombed them to hamper their progress. The missions, while successful, would see an incident which, while comic, was very nearly tragic.

For the first raid on 13 September, nine Halifaxes split into three groups. JP234 and JP183, flown by F/Os Simpson and Wheeler, and JP232 flown by W/O Fleckney targeted Tatoi Aerodrome in Athens.[38] Returning, the Halifaxes began their approach to Amendola in the darkness, following each other, awaiting permission to land.

Gunner Frank Pepper recalled, 'Just ahead of us we could hear W/O Fleckney calling for landing instructions but, before control could reply, an unknown aircraft cut across the centre of the circuit taking out one of the tailplane/rudders from Fleckney's Halifax.'

With his R/T still open, they could all hear Fleckney calling his crew to 'BAIL OUT – BAIL OUT!' It seemed the bomber was doomed. But Frank recalled, just seconds later, the pilot changed his tone and his message to his crew was, 'Relax you lot, it's going to be OK.' It seemed the Halifax might survive.

Frank's bomber landed and taxied to dispersal, next to Fleckney's aircraft. Both crews boarded the truck to take them back to camp, minus one crewman. They then found out what happened. 'The navigator,' Frank recalled, 'who sits on an escape hatch in a Halifax with his parachute stowed alongside him, had clipped on his chute, lifted the hatch and dropped through.'

Frank recalled, as they drove out onto the main road, one of the crew suddenly saying, 'Well, look who's thumbing a lift!'

There was the Navigator, F/O Facey, with, as Frank remembered, 'his 'chute rolled up and tucked underneath his arm.' Frank admitted, sadly, that, 'a great moment was lost by not having a camera to capture the look on the navigator's face as we hauled him over the tailboard and he realised he was with his own crew.'

The following day, at briefing, the CO stated, 'Gentlemen, as ably demonstrated last night by someone who shall remain nameless, there is absolutely nothing to be gained, with regards to an early breakfast, by jumping out of aeroplanes.'[39]

The squadron diary described events in more sober, but still startling, terms and paid tribute to the pilots' skill in averting catastrophe.

'What might have been a fatal flying accident was averted by the skill of pilots Flt Lt Moncrieffe and W/O WTM Fleckney. While coming in the circuit, waiting his turn to land, Flt Lt Moncrieffe suddenly saw another aircraft heading directly towards him. He at once pulled the nose up, missing a head-on crash by inches, but his bomb-bays sheared off a portion of the tail of W/O Fleckney's aircraft.

'This aircraft immediately fell away out of control, losing height very rapidly. W/O Fleckney managed to regain control, eventually landing it with no further damage. Flt Lt Moncrieffe's aircraft also landed safely.'

The diary added, 'In the confusion, when Fleckney's aircraft was out of control, F/O R Facey, navigator, lost his helmet and assumed that the order had been given to bail out. He duly donned his parachute and jumped, landing safely some four or five miles away from the aerodrome. On his descent his boots fell off and had a most unpleasant walk back to camp.'[40]

There were further raids that month in support of the 8th Army – all on a bridge near the frontline. The Allied assault on the Gothic Line, which 614 had supported in September, was bogged down.

German supplies and troops flowed across the pontoon bridge at San Benedetto over the River Po, now vital as most other bridges had been destroyed by Allied bombing. To stop that flow and help the 8th Army, the bridge became a priority target for 614. From 22 September to 4 October, it flew successive raids, often in terrible weather, against the heavily-defended bridge – yet they still managed to illuminate the target precisely with flares often landing on the bridge itself.

This excellence in pathfinding was recognised. 'From the PFF point of view, this month has been the squadron's most successful to date. The majority of targets have been of a tactical nature in support of the 8th Army. No aircraft were lost and target-marking was very accurate.'[41]

Throughout the rest of the autumn of 1944 the squadron would focus mainly on strategic targets – but December marked a shift and it operated frequently over Yugoslavia, both in dropping supplies and striking the retreating Germans.

In early October 1944, the Soviets entered Yugoslavia and moved to block the German withdrawal from Greece. They took Belgrade on 20 October and, by the onset of winter, the partisans controlled the east of the country.[42] The still-substantial Axis force was pushed into a corner in southern Yugoslavia, and partisans reported German columns retreating north-eastwards from Podgorica. The chance was there for 205 Group to pound them from the air.

On 3 December reports noted that the enemy, surrounded in Podgorica, was trying to escape through nearby Klopot. The tempo of operations increased and, between 15 and 21 December, 205 Group sent 316 sorties there with 614 Squadron at the forefront.[43]

On 16 December, six Liberators and five Halifaxes left Amendola to attack vehicles and troops around Mateševo. Les Sims recalled the dramatic mission. 'We'd had reports that troops and transport were pouring out in one continuous line on the only main road through the mountains. We were asked to verify, harass, survey the route, and report on the best way of slowing down the withdrawal.'

Deep in a narrow ravine, the road following a mountain stream made a sharp 'S' bend and, at the same time, crossed, via a bridge, the stream below. If this bridge could be put out of action, the column would be halted and partisans could attack. They took this recommendation back to Italy. Before going they left a calling card.

'We let down into a valley that was crowded with transport. Turning our .50 guns to port, the gunners left the entire line in flames.'[44]

On 18 December, eight Liberators and four Halifaxes left Foggia to bomb vehicles and troops from Bioče to Kolašin. The squadron's diary recorded how, 'thick concentrations were found stationary along the road. Many direct hits were claimed on motor transport and the road, and two aircraft strafed, destroying two large vehicles and damaging others.' Rarely had the squadron seen such direct results of its actions – in contrast to bombing from high level at night.

The next day, three Halifaxes and one Liberator returned to attack Mateševo. The target was the bridge that Les Sims' crew had spotted.[45]

He described how difficult it would be to destroy. The bridge lay at the bottom of a 'chimney-like ravine' but in an 'S' shaped slot which meant it was difficult to see. They made one pass and missed but, as there was little or no opposition, they made another pass. He recalled with satisfaction, 'The second time the load went down the chute.' They had got it. The bridge was destroyed.[46]

Partisans who surveyed the scene on 31 December confirmed the raids' success, reporting some 800 vehicles destroyed or abandoned between Podgorica and Kolašin due to air attacks.[47]

Whether allowing the enemy to escape or resupply, bridges continued to be an important target.

In the new year, the drive to destroy the Germans transport networks intensified. On 12 January, the Russians opened their winter offensive which would get them within 35 miles of Berlin by early February. Attacks on transport would be vital in stopping German reinforcements moving from Italy to the east, or to Western Europe.

The enemy's movements became more chaotic as the Allied air attacks told. Soldiers abandoned vehicles, others bought fuel on the black market to try and keep moving. Troops travelling from the Austrian border to Bologna would normally take twelve hours. This journey now took three to four weeks of changing, marching, hitch-hiking, waiting for long hours on roadsides or in tunnels.

By early February, the Germans resorted to ever-more desperate, although ingenious, ways to keep transport flowing. Temporary bridges were built under a few inches of water so they couldn't be seen from the air. Another ruse was the 'night-operational' bridge. A damaged structure would be repaired to a point, just short of completion, leaving one span missing so that aerial reconnaissance would class it as impassable. At night, a span would be dropped into place so traffic could pass.

So effective were strikes on marshalling yards, rail and road bridges that the enemy had to find another way to move their materials. Shipping was one of the few options they had left, and ports became the next targets.[48]

The bulk of these raids undertaken by 614 were carried out in February, and the Italian naval dockyard at Fiume (now Rijeka in Croatia) was attacked on 16 February, with Trieste naval dockyards hit the following day. But the next raid on a naval base, at Pola, on 21 February, would bring great reward, but at great cost.[49]

Tom Scotland recalled the morning of the mission. Flying alongside him would be fellow Australian pilot, Bill Neubeck. Scotland liked the recently-arrived pilot and they often played chess together. He remembered Neubeck's friendly, warm and smiling face that morning as they went for their briefing.

Pola was a base for U-boats, and Tom recalled that aerial photographs showed large supplies of German guns and ammunition being transported for storage in warehouses at its docks. For the daylight raid, they would fly along the Yugoslav coastline and turn north-west to attack the docks. In fine and clear weather they had breathtaking views from the flight deck of the rugged mountains of the Adriatic coast.[50]

Their smooth ride ran into problems. One Liberator's engine failed 45 minutes from the target, and a red TMB hung up after they were hit by flak. Others released TMBs, and bombing was reported to be well concentrated in the target area.[51]

Scotland recalled the severity of the fire directed at them and defences were 'wide awake' as they flew in. 'Black clouds of flak bursts hung in the sky and I realised that Altair was lined up in the sights of the gun batteries below.' The bursts came thick and fast, but exploded just ahead of them, so close that he wanted to put the brakes on, as if in a car. 'I dropped our speed back, hoping that the gunners would not detect the change, but they did. The shell bursts came nearer. 'Hang on, skipper,' came the bomb-aimer's voice, 'I can see the target. Left... leeeft.... left, left... steady... hold it there...' Tom steered the Liberator and watched as the exploding shells marked the sky. Out of the corner of his eye he caught sight of the other 614 Pathfinder Liberators 'arrayed across the sky as if in a cavalry charge'. They flew on together into the shellfire.[52]

That accurate anti-aircraft fire found its mark and one of them was hit. 'A Liberator near us gave of an ugly glow of bright flame and black smoke. In seconds, billowing flames engulfed the wings and then the whole craft,' Tom remembered.

The huge bomber slowed and lost height, then split into two parts, pieces coming away as fierce flames trailed into a long falling arc behind it. They saw parachutes come out, 'how many, it was difficult to tell; the sky seemed filled with smoke, flames, aircraft and two or three parachutes falling in amongst the chaos.'

They couldn't let the loss distract them from their mission. The target had to be marked. The target had to be hit. Ted Budd, the bomb-aimer called out, 'Hold steady... hold steady... That's good – hold it there.'

'Our bomb doors rumbled open and we flew on into the fury over the naval base at Pola, fiercely alive now with gunfire and exploding shells across the whole canopy of sky. Budd called, "Steady... steady... Bombs gone!" The TIs fell towards the docks. Thousands of feet below, they exploded. The aiming point would soon be a glow of green and red, marked out for the bombers following them. Navigator Bill Scott called out the course for home and, as the Liberator turned, a shout came out, "Oh skipper, look at this!" gunner Al Harper exclaimed. "We've got it, we've got it, there's a pillar of smoke shooting thousands of feet into the air. You ought to see it!" The ammunition stores had taken direct hits and had exploded.'

The crew looked on in amazement as the black smoke billowed out of the Pola docks.[53]

At the debriefing, Bill Neubeck and his crew were missing. The squadron diary told how the stricken bomber fell from the sky, and five parachutes were seen. Could the crew have survived? Most did not. Tom recalled that months later, the Red Cross gave the news that the bodies of Bill and his bomb-aimer were discovered in the wreckage at Pola.[54] Despite the report of five parachutes, of those who managed to escape the aircraft, only two would survive – A/Br F/O Gladstone and F/Eng F/Sgt Sugden.[55]

The remainder are all buried at Udine war cemetery in Italy. They were Pilot Flt Lt William Francis Neubeck; Navigator Flt Lt Harry Kenneth Mepham; WOP F/S Albert Edward Tyreman; Waist Gunner F/S Frederick Kendall; Sgt Christopher Wilkins; F/S Harry Gardner Lawley and Bomb-Aimer F/O Douglas Stuart James Gladstone.[56]

Following the raid against Udine in January, where he'd made his debut, Ted Budd would return on 7 March. However, this raid would have a very different outcome for him.

That day, ten aircraft flew to marshalling yards in north-eastern Italy, with five going to Cremona and five to Udine. Although there was dangerous weather en route to Cremona, with icing, there was a clear patch over the target with good visibility.

One Liberator, KG952, piloted by Les Sims, had to return early with engine trouble. It was descending to try and clear icing when the port outer engine ran away (following loss of control of the propeller and with it the risk that it would break away and hit the aircraft) and the air speed indicator stopped working. The propellers on the port outer engine were feathered and revs were increased on the remaining three engines. At that moment the rudders became very heavy and the aircraft spun to port. The pilot pulled out of this and the aircraft spun to starboard. This again was recovered from, but the stricken bomber then nearly went over on its back. By now it had lost 8,500 feet and Sims gave the order to bail out. Gunner Sgt David Nicholl and Bomb-Aimer F/S Ted Budd were the first to go, but the aircraft was then brought under control but still in a very steep dive. The order to bail out was cancelled and the aircraft was eased into a shallow dive and the controls became easier. It later landed safely on three engines.[57]

Tom Scotland was one of the crews sent to Cremona that night with a new trainee bomb-aimer after Ted Budd had graduated from operational training with them. They were sad to see him go. 'After eight attacks on enemy targets, he had learnt much from Ed, and we would miss such a man who had played his part well in our crew.'

They returned from Cremona to hear the shocking news. 'One of our planes had iced up and nearly crashed. Two crewmen were down in the Adriatic. One of them was our man Budd. But with night-time darkness, and a raging storm at sea, the search was being delayed till early morning light.'

He spoke to Les Sims who told of the terrible icing they'd experienced. 'The ice had built up so rapidly that the plane had gone out of control, spinning down. Believing the situation had become hopeless, he'd given the order for his crew to abandon the aircraft.'

The next morning, search aircraft flew over the previous evening's route. Tom and Ed waited for news but, with the storm continuing, the search aircraft had to stand down. He sadly recalled, 'Freezing conditions down on the ocean meant that Budd and Nicholl could not have lasted long. Ed and I departed for bed with heavy hearts.' He remembered it took them days before they could accept the twist of fate which had led to the loss.[58] They would be the last of the squadron to be killed on combat operations.

Sgt Nicholl's mother wrote to the squadron in her anguish asking to know how her son had died and would his body be found. The reply said that his parachute, at least, would have had time to open, but they could not tell much more.[59]

Just days before, on 4 March, Ted Budd had flown with Tom Scotland in Altair, KG945, when they'd had a miraculous escape. Nine Liberators flew to mark Casarsa's marshalling yards in northern Italy. Scotland's Liberator had just dropped their red TMBs when, as navigator Bill Scott recalled, 'there was a hell of a bang and a flash between the tail-plane and the wing on the port side. The aircraft shuddered, and I thought it was "flak" but it sounded different.' Communication was lost between the flight deck and the rear of the aircraft where the wireless operator was manning a waist-gun.

As they left the target area wireless operator, 'Shorty' Weekes, pushed up the hatch at the front of the bomb-bay (leading to the flight deck) and shouted, 'Everyone alright here?' Bill said, 'I almost laughed because he was soused in oil from head to foot. Tommy Scotland said the flash had been a photo-flare.' But they later discovered that it was two incendiary bombs which had fallen through their aircraft and left through the floor, leaving gaping holes.

One of bombs, Bill recalled, severed a hydraulic pipe, spraying 'Shorty' with oil. The loss of hydraulic fluid was going to cause them problems and, as they neared Amendola, the landing gear would not lower. First the port, but not the starboard, wheel would drop. After retrying the opposite would happen.

Scotland warned Amendola that they had a problem. Eventually, both wheels dropped, but there was no indication that they were locked. The nose wheel, meanwhile, was still retracted, so bomb-aimer Ed Riley went forward and kicked it until it went down. Bill recalled, 'There was still no indication that the landing gear had locked down in position, and of course the hydraulic fluid was gone.'

With suspect landing gear and no hydraulic fluid to power the brakes, they prepared for the worst. 'Shorty' was told to strap a parachute to a waist-gun position, and prepare to pull the rip cord if the brakes failed in the hope that it might help slow the aircraft. The lumbering bomber touched down. Weeks was braced ready to deploy the makeshift brake 'chute. Tom Scotland applied the brakes and hoped.

Bill remembered, 'Luckily, there was just enough pressure trapped in the oil circuit to apply the brakes and get the aircraft to stop before running out of runway. All the airfield fire engines and ambulances were standing by, fearing the worst – happily it was otherwise.'

The crew clambered out and saw what the bombs had done. They were amazed to see other incendiary bombs had lodged in the wings. Miraculously, they had not ignited.

Following the incident, 205 Group decided that in future they would send aircraft carrying incendiaries at the same height as the visual markers (i.e. about 6,000 feet, rather than the 10,000 to 12,000 feet).[60]

By April, the net around Kesselring's German forces was tightening in northern Italy. The 8th Army on the north-east coast of Italy and the US 5th Army on the west coast were closing in. Perhaps this was the endgame, but the remaining Axis forces still had to be destroyed, captured or neutralised.

That month 205 Group would launch 1,251 sorties, half of which (629 sorties) were in support of ground forces.[61]

Again, 614 Squadron would play a big role, for which they'd get wide recognition. Between 9 and 19 April, the squadron carried out raids daily to support advancing troops. To coincide with the launch of the 8th Army offensive on 9 April, 205 Group was tasked with destroying concentrations of enemy troops around the Santerno River. It would need high precision because the enemy would be just 2,000 yards ahead of the advancing Allied troops. A plan was drawn up where the Army would fire red marker shells which 614

Squadron would drop their target indicators on. Owing to extreme accuracy being required, no second runs were to be allowed.

Barry native Fred Seal, a bomb-aimer flying Liberators with 70 Squadron as part of 205 Group based at Foggia, frequently bombed targets marked by 614. He recalled briefings before the raids left crews in no doubt that great precision would be needed to ensure that friendly troops were not hit.[62]

On the night of 9 April, 614 sent eleven Liberators and the diary told how red star shells were fired by the Army as planned. The lead Pathfinder (or Master Bomber) directed visual markers to mark briefed points, dropping them at 03.58 and 03.59. Timing was that tight, but it worked.[63] The 205 Group report told how an 'exceptional concentration' was obtained on the target code-named 'Pig'. Overall, not a single bomb fell outside the area laid down, and 'utter devastation' was caused along the river defences, which enabled the ground forces to establish their first bridgehead the next day with 'very little opposition'. There was less success on target 'Whistle', however, which was not illuminated by ground forces.[64]

The next three nights saw more support for the 8th Army around Bastia. No. 205 Group was told to 'destroy everything in the defined area' just 5,000 yards in front of British troops. On 11 April, again, split-second timing was essential and flares were dropped from 20.53.07 to 20.56.00 with red TMBs dropped at 20.57.02 and 20.57.07.

There was praise, too, for the now Wg Cdr, LWJ Moncrieff, who acted as Master Bomber and directed the 'extremely well-concentrated attack' by 80 Liberators. He would later be awarded the DFC for the role he played that night.[65]

March also saw the DFC go to Flt Lt CH Muggeridge, A/ Sqn Ldr A Dunning, Flt Lt WA Irwin and Flt Lt W Simpson.

As the Germans retreated, so 614 helped to harry them. More devastation was wreaked on 17 April when 205 Group returned to Portomaggiore. Eight Liberators from 614 marked targets 'to great effect' and photographs showed severe damage to the area, with through lines blocked, a road and footbridge demolished, roads severely cratered, and such extensive devastation that all road traffic would be forced to make a detour.[66]

There was recognition for another 614 Squadron Master Bomber following these raids. Wg Cdr EBR Lockwood MBE, acting CO, was awarded the Distinguished Service Order for co-ordinating the attacks. His citation read, 'This officer directed a full-scale night attack against the strongly held enemy line, south-east of Bologna, and only 2,000 yards in front of 8th Army troops. His careful planning and masterful control of the attack resulted in a highly-successful bombing concentration and gave valuable support to the ground forces.'[67]

No. 205 Group received a letter of congratulations from the Allied Supreme Commander, Mediterranean, General Alexander, endorsed by the 8th Army Commander, General MacCreery.[68]

He wrote, 'Most grateful for the heavy and continuous support you are giving to the 8th Army. Your accurate and devastating attacks are causing great disruption to the enemy. Your bombing of the Santerno defences played a most important part in forcing this river obstacle. The forward troops who had a grandstand view of the bombing were reported to be "very enthusiastic".'[69]

Now the end was in sight. The last raid was on Freilassing's marshalling yards in Austria, which was believed to be part of a last stand in the Alps. Hopes it would supply this 'Nazi Redoubt' were dashed by the night attack on 25 April, which destroyed depots and blocked the whole yard.[70]

Despite the great success in helping the advance of Allied troops, the month would end with tragedy. In February, 205 Group investigated using the Mosquitoes as Pathfinders, and by March the first of them had been delivered. One was lost on 16 March, when LR470 crashed on the coast north of Bari on a training flight. Pilot F/O Ronald Kingston Dorrington and Navigator P/O Charles Henry Boston were killed. Then, with the end in sight, another - LR442 was lost on 30 April. Pilot F/O Brian Charles Robinson, RNZAF, and Navigator F/S David Idris Mutton were killed when it crashed at Capodichino. Yet another, KB531, was lost on 25 June. Pilot Flt Lt Walter Edward Jay and Navigator P/O Henry Claude Willis were both killed when it crashed off Malta en route to Luqa.

The squadron did not operate again until 7 May, when two Liberators took off, not to drop target-markers or bombs, but leaflets. Leaving at 19.05 they dropped their cargo around Graz in Austria, returning at 00.40 on 8 May. It would be the last time a 614 aircraft would fly on war operations.

Later that day the Allies formally accepted the unconditional surrender of Nazi Germany and brought an end to the war in Europe.

The news was greeted with some festivities but also much new work. 'All Messes held celebrations and a good time was had by all.' Messages of congratulation flowed in, including from the King who sent 'heartfelt congratulations on the... triumphant end to your long and arduous campaign in Italy.' The CO addressed a packed cinema and told them of the new task it now had. VE Day would be a normal working day but there would be 48-hour passes for all 'at the first opportunity'.[71]

What was that new task? The collapse of the enemy in northern Italy and the 8th Army's rapid move into Austria had thrown massive strains on thin lines of communication, and No. 205 Group were to transport supplies to forward units until Trieste port became available.

Also, although the war against Germany had ended, a dispute with a former ally now threatened to escalate.

Partisans of the Yugoslav National Army had advanced rapidly around Trieste and into Austria, refusing to return to agreed demarcation lines. They held full control of Trieste until 12 June and it seemed that the Allies might have to fight to push them back.[72]

The first of these missions came on VE Day itself, when 13 aircraft delivered 395 boxes of 'Compo' rations and 1,008 tins of petrol to Rivolto Airfield, near Udine.[73]

The schedule was hectic and squadrons had to make two deliveries per day. Some of the grass airfields became rutted and unusable after the heavily-laden Liberators landed on them. One aircraft of No. 178 Squadron crashed, killing the pilot. But, by the end of this operation, on 25 May, the Group had flown 1,225 sorties and over 4,000 tons of rations of all sorts, petrol and oil were transported.[74]

'The cessation of hostilities,' the 205 Group diarist explained, 'instead of reducing the Group's activities, had the effect of increasing the number of sorties during the month to a figure approximately 150 higher than the average for the previous six months.'[75]

The squadron also had a happy duty – flying freed POWs back to the UK. As prisoners flooded into collection points across Europe, they needed to be taken home quickly. So RAF bombers flew them home in 'Operation Exodus'.[76]

To do this, benches were built into the Liberators' bomb-bays; armour-plating and bomb-sights were removed, and only the most reliable machines were used. Crews made flights without passengers to familiarize themselves with the route to transport their precious cargo.[77]

The first missions were completed on 15 May. Missions ramped up towards the end of the month by which time the squadron had made 48 trips and flown 1,203 Allied POWs home.

For Les Sims there was real satisfaction in taking the prisoners home. He called it 'a final pathfinding duty, this time the "target" was for the best reason of all, home and peace.'[78]

Early in the month, the CO Group Cpt JS Laird left the squadron and was replaced by Wg Cdr EBR Lockwood.

There were also hints of being sent to fight in the Far East and the Group was gathering training material for the conversion to the Lancaster and films on Japanese aircraft and tactics.[79]

Many crews were going home. Throughout June and July, with the departure of the Australians and South Africans, the squadron was becoming less and less like a League of Nations. Before Tom Scotland and 'Shorty' Weeks left for Australia, the crew all went into Foggia town together. The weary crew smiled for the camera as they were together for the last time. Tom felt proud of them all.

Early in May 1945, work began in preparing men for life after demobilisation and the educational and vocational training scheme began. This included debates on current affairs which aimed to expand awareness of the civilian world they'd soon be re-entering.

The first debate was 'Parliament – the Picture Today'. There were lectures on 'Release and Resettlement', and on 'National Finance, Leisure and Trade Unions'. There clearly was a thirst for discussion with a unit debating society formed which pondered the motion that, 'The system of private ownership of the means of production has justified its continuance, and a system of public ownership is undesirable.' The motion was defeated 'by an overwhelming majority'. By July's end, 150 personnel were getting five hours education per week, and training was given in subjects such as motor transport, carpentry and music. There were also cultural excursions to Italian cities.

The end of the war also brought many honours for the squadron. On 12 June there were DFCs for F/O Hugh Fleckney, F/O Tom Scotland, P/O GHR Slade and F/O TW Parker. There were Mentions in Desptaches for Sgt AHMWilson, LAC M Mead and posthumously for F/S Ted Budd.

Meanwhile, despite the end of hostilities, improvements to the domestic camp continued. However, by late July, the diary told, 'in view of the impending move, all further work has been suspended'. Change was looming and it came swiftly. On 27 July, the squadron completed its final operational flight when a Liberator delivered 25 SAAF personnel to Cairo. At midnight, 614 Squadron ceased to exist.

By the stroke of a pen, just as swift as the one which had given the squadron life again in March 1944, it was again taken away. For the second time, No. 614 (County of Glamorgan) Squadron was disbanded. But unlike in January 1944, when its men were scattered – they would stay where they were and take another squadron's number.

As the squadron diary put it, 'On the 27th of July 1945, the Squadron number was changed to No. 214 Squadron. The squadron would be moving to Ein Shemer, Palestine, on or about 15 August 1945.'[80]

Their time in Italy had been 614's longest and most intensive in the war. It operated there for 14 months and left most of its fallen in central and Eastern Europe and Italy.

George Harper, along with all others in 614, became a member of 214 Squadron. By November he was at RAF Fayid in Egypt, where they were ferrying British personnel back to the UK. He gave a moving postscript to 614 Squadron's time in Italy. George and his crew would be flying back home to be demobbed but were told they'd have some passengers – 24 WAAFs. They were soon on course and heading for Bari, their overnight stop. George recalled, 'The weather was fine and sunny and our "cargo" were enjoying themselves. We were really on our way at last.'

But they were not there yet. The operation involved six aircraft, five carrying passengers and one carrying mail but as he explained, 'As we flew on towards Italy, cloud started to build.' They couldn't make contact with one of the aircraft. When they landed, they found that the aircraft carrying the mail was missing. 'We learned that it had crashed in the mountains of Crete, with no survivors.' He added, 'We were very shocked and saddened for something like that to happen on our homeward journey.'

When the weather cleared they took off on the final leg of their journey. Crossing the Channel, 'there was the English countryside beneath us' – it was an emotional moment.

They landed at Glatton in Cambridgeshire and taxied to their dispersal point, saying goodbye to their passengers. But it became clear their journey was not done yet. They were told they had to fly to St Mawgan in Cornwall. There was confusion there, too, and they were sent to Lichfield. 'Once more we took to the air,' George recalled, and on arrival at Lichfield, it was clear this was the end of the road. Aircraft were parked wing tip to wing tip to await disposal. They landed and, once parked, cut their engines. 'All was silent,' he remembered. 'We had reached journey's end.'

They all picked a keep-sake from their Liberator, their service, their war. George took an oxygen mask bag, which he put to good use. 'It made a useful shoe brush bag.'

They took a train to a dispersal camp at Lytham St Annes, near Blackpool, and had a final night out using the British money they'd been given and went their separate ways the following day.[81]

This one final act, perhaps, encapsulated much of the war experience, from the unexpected 'cargo', the waiting, the peril and terrible loss and, then, the emotion of seeing home again. Finally, after some delay and confusion, they had a drink together.

CHAPTER 10

Llandow (1946–1957)

After the squadron was disbanded in 1945, there was no swift phoenix-like return from the ashes as there was in the spring of 1944. Peace meant Britain's armed forces were scaled back and millions were demobilised.

But before the war had even finished, there were discussions on the future of the Auxiliary Air Force which would have a profound effect on 614's future. It was decided that the AAF squadrons would be reformed with most operating fighters, and recruiting was to begin in June 1946.[1]

At 15.30 on 11 June 1946, the first member of the reformed 614 Squadron reported to RAF Llandow for duty. He was Flt Lt GA Martin DFC & Bar, and would be the sole member of the squadron for a few days. He was soon joined at the airfield 20 miles west of Cardiff by two officers and 14 other ranks (ORs).

The squadron was to be brought up to full strength (17 officers, 19 SNCOs and 133 ORs) by 1947, by which time it would have two flights – one for training with two Harvard two-seat trainers and four Spitfires, and an operations flight with nine further Spitfires. The Spitfires would be MK XVIs – a much more powerful version of the type which fought in the Battle of Britain. Spitfires were no strangers to Llandow and many were based there during the war. The link is perhaps best-known via the poem, 'High Flight', written by P/O John Gillespie Magee Jr while stationed at Llandow in August 1941, after he flew his Spitfire to 33,000 feet. He was inspired by what he experienced to write his poem.

The CO had not yet been appointed, but there was a familiar face who would play a leading role in the development of the post-war squadron. Richard Cadman, the squadron's first CO, remained as Honorary Air Commodore and he paid the first of many visits to the squadron that month.

Milestones soon came. Their first aircraft, two Harvards, arrived in September and, by the end of October, two more pillars for the squadron's future were in place. On 29 October, there was a visit to the site of the squadron's 'town' HQ in Cardiff. Having a centre for recruitment in the city gave 614 a presence that isolated Llandow could not. Although in the Ely area of Cardiff, it was known as RAF Llandaff to avoid confusion with Ely, Cambridgeshire.

Two days later the CO was appointed. Another familiar name to the squadron, he was Sqn Ldr WH 'Bill' Irving. A wartime veteran of 614 and from Cwmbran, he'd first joined the squadron in 1941. He'd served with distinction in North Africa, particularly when he saved the lives of his crew and himself during a disastrous raid on an airfield in Tunisia in 1942. Leaving 614, he went on to serve in Bomber Command, notching up over 60 missions and a DFC. There could be no more fitting or fine officer to lead the squadron.

There were more echoes of the past in November when recruitment began with an open day at the Drill Hall, Dumfries Place, Cardiff, where its first recruitment event was held in 1937.

The new year of 1947 began brightly with fighters added to the squadron's strength – now three Spitfires joining the two Harvards. On 22 January enlistment of Auxiliary personnel began with twelve recruits expected, but only four turned up. But it was noted that they were 'particularly keen types'.

By April, the first flying training for recruits began and they were given some unfamiliar ranks which had recently been introduced for non-commissioned aircrew.[2]

NCO ranks from Warrant Officer to Sergeant disappeared, to be replaced by five new ranks, Aircrew 1 to 4, with a new rank corresponding to Warrant Officer, Master Aircrew.

The insignia on these new badges was of laurels and stars, but the style was unpopular, and was cynically known as the 'star and garter' badge. Air Ministry policy bowed to popular distaste, and in August 1950 the aircrew ranks 1–4 were abolished, to be replaced by the former flight sergeant and sergeant titles. Only the Master Aircrew badge and rank was retained as aircrew equivalent to ground branch Warrant Officer, and it remains to this today.[3]

A big part of the life of an Auxiliary squadron was the annual camp and the first since reforming would be at Llandow. The camp began on 23 August with the aim to fly 250 hours, with a 08.30 start each day and a finish at 17.00.

Once flying began again, one of the exercises was to practise air interception – a key role as a day-fighter unit. The Spitfires' interceptors would be 'scrambled' and would hunt the Harvard 'bombers.' The Harvards were airborne at 10.00 and, starting from near Brecon, carried out a dummy dive-bombing raids on Port Talbot. The intention was that a section of Spitfires would reach the Harvards before they got to Port Talbot and be intercepted on their way back to Brecon over Merthyr – meaning their raid had been successful.

October saw the squadron's first accident when a Spitfire belly-landed after Flt Lt Martin was unable to lower his undercarriage. Two months later, there would be a far more serious crash, with fatal consequences. The flying for the December training weekend had low-flying formation of P/O DE Morgan and two Pilots 2, Mitchell and Hodson. After completing the manoeuvres, the three were turning before heading home when Pilot 2 Hodson collided with his leader, P/O Morgan. Hodson parachuted to safety, but P/O Morgan was killed when his aircraft, Spitfire RW355, crashed at Long Sutton in Somerset, having lost its tail in the collision.

Flt Lt Rich flew to Yeovilton to collect Sgt Hodson, and P/O David Edward Morgan was buried four days later. All squadron personnel were at the service at Tynycoed Congregational Chapel, Abercrave, to say goodbye to their fallen comrade.[4]

Almost all would have served in wartime, but death's sting would hurt just as much despite coming in peaceful south Wales rather than war-torn Europe. The year ended with an honour from the King. On 16 December, the Auxiliary Air Force was granted the prefix 'Royal', in recognition of its service during the war.[5] This success, Sqn Ldr Irving admitted, was bittersweet, adding the year was marred by P/O Morgan's death.

The new year saw further progress. An exercise with other Auxiliary fighter squadrons was planned, and ground-firing training started with help from RAF Chivenor, who would staff the range at Steart Flats in Somerset.

However, that bright prospect was soon dimmed when news came of another fatal crash involving a 614 pilot. On 17 April, P/O Howard Hughes Evans reported for flying duty at 10.00 and took off for a flight which should have lasted 30 minutes. After 25 minutes, Llandow tried to radio him after the weather deteriorated. Contact could not be made and reports came in of an aircraft seen to emerge from cloud in a dive and then plunge into the sea near Barry Docks. Air Commodore Cadman and Flt Lt Arthur went to Barry to collect and identify debris found at the scene of the crash. These were found to belong to P/O Evans.

The Air Commodore, along with Flt Lts Spurdens and Martin, visited the pilot's wife and parents to offer the condolences of the squadron. 'P/O Evans was a very popular member of the squadron and the tragedy has been a great loss to the unit,' the diary noted.

As always, in peace and in war, squadron life had to continue in the face of these tragedies, and in the wake of P/O Hughes' death there were signs of a brightening future for the squadron.

On 20 June, news was received that 614's Spitfire XVIs were to go. This version of the famous fighter was powerful. With its Packard-built 1,500hp Rolls-Royce Merlin engine, it was also the last major version of the Spitfire to be powered by the Merlin. Former RAF Battle of Britain memorial flight pilot Cliff Spink praised the delicacy of control the powerful Spitfire Mk XVI had. 'It was so light to handle, perfectly trimmed out and "hands off" – beautifully balanced.' A feature common to all marks of the Spitfire was that the cockpit was snug. 'The cockpit itself was tight, but that had its benefits – you felt as though you strapped on the Spitfire and that gave you an immense feeling of being at one with the aeroplane.'[6] With two 20mm Hispano cannon and two .50 caliber Browning machine-guns, it packed a powerful punch, yet, its replacement with 614 Squadron was more powerful still.

Considered by many to be the ultimate Spitfire, the F Mk.22 shared only a passing resemblance to its early Spitfire ancestors and was powered by the far more powerful Rolls-Royce Griffon. This resulted in some changes to the airframe, including an increase of the tail surfaces. Like the Mk XVI, it had a bubble canopy and cut down rear fuselage. The wing area was also increased, as was its stiffness, thus greatly improving handling. Armament was superior with 4 x 20mm cannon and no machine-guns. Due to the large five-bladed propeller, the length of the undercarriage had to be increased. When the first one was delivered on 26 July, many said how small the Mk XVIs looked compared to them.

The first annual camp away from Llandow was confirmed

for Horsham Saint Faith, in Norfolk, between 14 and 28 August. The newer machines were not supposed to be delivered until after then, but it was believed one or two might be available.

After an air-test, Flt Lt Martin found that, 'while being faster than the Mk XVIs,' they were 'not so pleasant to fly and are certainly not comparable with them for Auxiliary flying'. The groundcrews would also struggle and it was felt, 'The maintenance on these aircraft is going to prove more difficult.'

The camp began with a hair-raising introduction to the more powerful Spitfire. On 14 August, Pilot 1 Graham Crocker was bringing one of the new machines in when it 'ballooned' while landing, stalled, and a wing tip touched the runway. The diary told how, although the incident was due to pilot error he corrected it extremely well, 'preventing serious damage to the aircraft'.

The camp was judged to be 'a huge success', with 341 flying hours notched up. Some of those stationed in Norfolk enjoyed meeting the visiting Welsh squadron, too.[7]

Ron Ridout was an RAF regular from south Wales who was based at Horsham Saint Faith in 1948. He was later posted to St Athan, but when he left the RAF, so impressed was he by 614 that he joined them as a reservist in 1954.

He was a fireman at the Norfolk station when 614 came for their camp and recalled, 'I was perched on top of a crash tender by the control tower watching the squadron arrive in their Spits, followed by two Harvards. I, being the only "Taffy" on the fire section, was thrilled for that fortnight to mingle with so many lads from home.'[8]

When the squadron returned, there was preparation for 'Exercise Dagger'. The squadron would be based at West Malling and, for the first time, it would be operating with large numbers of regular RAF units and aircraft of the USAF stationed in the UK.[9]

Taking place between 3 and 7 September, it involved all Commands of the RAF, the USAF, the British Army's Anti-Aircraft Command and the Royal Observer Corps.[10]

The aircrew from 614 were defending interceptors and the pace was unrelenting. By 5 September, 614 crews were on stand-by from 04.00 and were scrambled repeatedly. One of 614's Spitfires managed to get among a stream of B-29 bombers and attack. They were, however, unable to stop twin-engined RAF Hornet fighters, playing attackers, and 'bombing' West Malling with toilet paper, 'suitably inscribed'.

The exercise gave lessons which would have a profound effect on 614. It became apparent that the Auxiliary squadrons had been 'wired up' wrongly. When they had first been re-established, it seemed logical to place them in Reserve Command along with the RAFVR and University Air Squadrons. This didn't help in standardising training and 'Operation Dagger' showed that Auxiliary squadrons and fighter control units were working to procedures that differed from those used by Fighter Command. Operational control of the Auxiliary air defence units was transferred to Fighter Command in 1949.[11]

The pace of tasks continued after the exercise had ended. There were the Battle of Britain displays to prepare for and perform. On 9 September, four Spitfires flew in formation over the beach at Port Talbot and one Harvard did the same two days later. A single Spitfire displayed over Penarth promenade, and there were displays at St Athan and Fairwood Common in Swansea.

The diary told how, at this time, the last of the older Spitfires were disposed of and the squadron was 'sorry to lose the Mark XVIs'.

As ever, recruitment was a priority and the Air Commodore suggested a novel way of spreading the word. They would drop leaflets from aircraft, with the first 'sortie' being over

Mountain Ash on 11 October where a mobile recruiting team opened their campaign. Three days later they did the same over Pontypridd.

On 15 November the squadron made a flypast over Cathays Park in Cardiff in honour of the birth of a son, Charles, to HRH Princess Elizabeth which was announced that day.

Aside from Battle of Britain displays, much of 1949 was taken up with exercises. July saw 'Exercise Foil' at West Raynham in Norfolk, the airfield 614 had flown their first missions during the war, only seven years before. Weary pilots carried out constant scrambles and, at the end of the exercise, it was admitted, 'Saturday night sortie was up to standard and various local hostelries were visited.'

In September the squadron joined the multinational 'Exercise Bulldog', organised by the Western Union, a precursor of NATO – involving the RAF, USAF and the Netherlands Air Force in simulating bomber attacks and air defence of the UK.

On 24 September, a lively exercise produced an equally lively night out afterwards. 'F/O Morgan and P2 Crocker caught two Hornets who were attacking Llandow at low level; our pilots dropped out of the cloud at 5,000 feet and fastened on the tail of the "Rats" [enemy, the interceptors were "Terriers"], blowing them out of the sky in front of the crowd at control to much applause. A party in the mess that night was enjoyed by all with many Welsh songs rendered.'

Other changes that month saw neighbouring unit, No. 501 visit in their new Vampire jets from Filton – the diary noted 'they were too fast for our Spitfire 22s'.

In January 1950, incoming visitors included 11 Group electrical and armaments officers, and 614 pilots entertained them at the Globe in Llantwit Major, the evening concluding with 'a game of rugby in the Mess'. The squadron took its first tentative steps into the jet age when its first Vampire Mk1 was allocated to the squadron on 6 February.

Just weeks later, squadron life was turned upside down – 12 March would become one of the squadron's blackest days.[12]

The diary paints a picture of what began as an enjoyable day – to one of horror. Some of the squadron's aircraft flew in the afternoon, watched by large crowds of civilians who had come to see the return of a chartered airliner bringing rugby supporters from the Wales v Ireland Five Nations game.

In a thrilling game, Wales won the Triple Crown with a last-minute try. But 80 of the supporters, who had cheered them to that win less than 24 hours earlier, would die in what was then the world's worst air crash.

The Avro Tudor airliner, called *Star Girl*, carried fans from rugby clubs and pubs across south Wales. The aircraft had been initially booked for 72 passengers, but extra seating had been installed. Eyewitnesses recalled how the Avro Tudor approached Llandow very low, with the undercarriage down. Increasing power to gain height, *Star Girl* then rose steeply but stalled and plummeted towards the ground. The right wing tip hit the ground, followed by the nose and left wing, which broke from the fuselage as it hit the ground. The plane spun around and finally came to a rest near in a field near the village of Sigingstone.

Only three of the 83 onboard survived. Two of them were sitting in extra seats bolted in at the rear and walked away unaided, while a third man, who was in the toilet and was knocked unconscious at the time of the crash, survived but was in hospital for four months. Eight more survivors of the initial impact died later in hospital, bringing the final toll to 80.

A party from the Greenhouse Inn in Llantarnam, near Cwmbran, was also onboard and the death of one of them would have great significance for 614 Squadron. The trip was arranged by the CO, Sqn Ldr Bill Irving, who was arriving back at his squadron's home airfield.[13]

He'd been shot down in North Africa, survived Dunkirk, and flown Lancaster bombers on more than 60 missions.

The airmen of 614 went immediately to the scene. Under the direction of Flt Lt Verney, the engineering officer, they carried out 'magnificent work in extricating the badly-injured passengers, most of whom were dead or beyond aid, and who died within a few minutes.' Squadron personnel stayed at the scene until all bodies had been removed. They had 'the unhappy task of bringing out the body of our Squadron Commander, Sqn Ldr WH Irving DFC, who had been a passenger on the Tudor.'

In an unhappy coincidence where strands of Pathfinder history met, the Tudor was flown by Fairflight, the airline owned by former Air Vice Marshal and founder of the Pathfinder Force, Don Bennett. Bennett was also one of the first visitors to survey the crash site.[14]

Flt Lt Smythe and squadron staff were kept busy answering telephone calls about the accident as the station became the centre of press activity. He remained on duty to help contact squadron members as preparations were made for the CO's funeral. Held on Friday, 17 March, it took place at St Hilda's church in Griffithstown, near Cwmbran, and all squadron officers attended. After the service, his body was taken to Pontypridd for cremation and the ashes were brought back to Llandow by Flt Lt West-Jones to be scattered from a 614 Squadron Harvard. The simple ceremony was watched by Mrs Irving and other family members. Three Spitfires, flown by F/O Morgan, P2 Croker and P2 Jones, staged a flypast as a last salute.[15]

For the third time, the squadron had to recover from tragedy and fulfil their duty. In April, that duty involved familiarising themselves with the world of jet flight.

On 8 April, a Sqn Ldr Horsley arrived in a two-seater Meteor T7, and spent the day giving familiarisation flights in the twin-engined jet to squadron pilots. By the end of the month, 614's first jets, Meteor T7s, were operational.

Next month, the jet journey continued when conversion from Spitfire Mk 22s to Vampires began. At Llandow, two Hastings transports arrived for the airlift to the annual camp on the German island of Sylt. Having settled in on the island's airfield, training began in earnest. 'Today was fine,' the diary told, 'and air-firing details were off to an early start'. Re-arming between sorties was carried out with 'remarkable rapidity' by the armourers, who worked 'immensely hard' throughout the whole of camp. Conversion onto the Meteors also began with F/O Morgan becoming the first 614 pilot to go solo in a jet.

The first weekend of the camp was free and most took advantage of the brilliant sunshine to 'indulge in bathing' on the island. A low-lying island, Sylt's sandy beaches was a popular German resort, especially with nudists – as many goggle-eyed 614 personnel discovered.

August saw two notable milestones. Firstly, the new CO, Sqn Ldr KH 'Sam' McHardy took over. A winner of a DSO, DFC & Bar and Croix de Guerre, it was he who, on 12 August, made the squadron's first Vampire flight, despite low cloud and rain.[16]

The Vampire was quickly pressed into service and, later that month, a Vampire and a Meteor gave an aerobatic display over RAF Fairwood Common on the Gower for the Swansea Air Pageant. In September, 614 Vampires gave a formation displays at Penarth, Pembroke Dock and St Athan as part of the Battle of Britain commemorations.

First flown in 1943, the Vampire didn't enter service until the early summer of 1946. It had an unorthodox arrangement of twin rear booms mounted behind a moulded, egg-shaped fuselage, the front of which was made of plywood and balsawood, with the remainder being made of metal and housed a single DH Goblin turbojet engine. Like the Spitfire F2 Mk22, it was armed with 4 x 20mm Hispano cannon. But its maximum speed was 548mph, with a service ceiling of 42,800 feet, meaning it could outperform its predecessor – in some ways at least.

Fellow Auxiliary pilot Tom Moulson, with No. 601 (County of London) Squadron, recalled they replaced their 'beloved but obsolete' Spitfires with Vampires at about the same time as 614, in 1950. 'They could fly faster and higher than the Spitfires and were eaier to maintain but not much else.' The fuselage sat low to the ground so that on start-up, an airman had to stand-by with a fire extinguisher if it was carried out on grass, lest the engine shoot out flame and set fire to the airfield.

He remembered that flying the Vampire could be unsettling at first. 'Lacking the grip of a propeller, there was no immediate push in the back on take-off and it seemed that the plane would never leave the ground. Unlike the Spitfire and Hurricane there was no bulky engine in front blocking out vision, which was helpful, and no swing on take-off from propeller torque which, although this too was helpful, somewhat diminished the satisfactioni of taking off. Close formation was a little tricky at first because, as with all jets, engine response to movements of the throttle were delayed without the immediate pull or drag of a propeller, and one had to anticipate more. Like all jets it seemed heartless and less a steed than a machine. But, and this seemed significant, the squadron had moved into the Jet Age.'[17]

'Exercise Emporer', in October, was a large-scale air defence exercise which saw the squadron fly to West Raynham with Vampires for the first time. As the jet-engined future of the squadron roared on, so the piston-engined past of 614 was disappearing and, by October, the Spitfires were being disposed of. But, in a nod to their prop-driven roots, a Vampire made a flypast over Cardiff Airport (formerly known as RAF Pengam Moors, where the squadron was formed in 1937) where the new ATC headquarters was being opened. By the end of the month, all pilots had converted to jets.

On 29 January 1951, in response to the outbreak of the Korean War, Prime Minister Clement Attlee announced plans to improve military preparedness and speed up British rearmament. Proposals included recalling reservists for training.[18]

Eventually, all 20 RAuxAF fighter squadrons were called up for three months' full-time service in home defence in place of regular squadrons earmarked for deployment to Korea. In the event, RAF fighter squadrons were not needed in Korea, but the RAuxAF squadrons were retained for intensive refresher training at their home bases.[19]

Someone who joined 614 at this time would have their training interrupted by the 'call-up'. Geoff Watkins was an Army glider pilot who had left the Army in 1949. He worked as a draughtsman with Rhymney Engineering of Cardiff, and decided to join the reserves as a pilot.

He started training with the squadron in January 1951 and his first flight was with 'Jasper' Jones on 13 January in a Meteor MkT7. He explained that when pilots flew to practise instrument flying, both sides of the canopy were smeared with Brasso, so that the only clear portion of the canopy was the front part. 'When Jasper had finished this part of the exercise, he showed me how to loop a Meteor.'

After a few flights with Sgts Player and Weaver in a Harvard, he went for another familiarisation flight in the Meteor MkT7 with Flt Lt Chappell. Following more trips in a Harvard, he was told he would be ready to go solo, but unfortunately it was not to be as, around this time, all flying members of 614 were called up for their three months continuous training.[20]

Some aircrew were detached to RAF Acklington, in Northumberland, as part of the call-up.[21] This time was not without incident. Vampire VF336 was written-off on 21 May after it made a force-landing on a beach nearby after running out of fuel in bad weather.[22] The squadron would lose another one that month, but this time without a soft landing.

Vampire VV196, flown by Pilot 2 Eddie Campbell from Deri, Bargoed, was training over RAF Pembrey. He was pulling out of a steep dive on the range when the jet rolled three times before striking the sea in a shallow dive. He did not survive.[23]

Barri Hurford Jones worked in the orderly room at Llandow and was on duty when Sgt Campbell died. He remembered the personal effects of the new pilot being brought in following his death. 'He was with the RAuxF for just a few weeks and his identity papers, issued by us not long before, were all soaked in sea water when they were returned.' Sgt Campbell was recorded as being a 'new' pilot on 11 March and, as a P3, had a flight in a Harvard with F/O Mitchell. Soon after, he had his first solo flight in a Spitfire. It was believed that the Vampire entered a high-speed stall because Sgt Campbell made too violent a recovery from the initial dive.[24]

That summer 614 went to Tangmere – the squadron's first all-jet camp now that all had gone solo on the Vampire. They were visited by the Lord Mayor of Cardiff who was given a display of close-formation flying. A party was held in Chichester to mark the end of their stay in Sussex.

The coming month would also see some of those Vampires meet untimely ends.[25]

Tom Scott arrived at RAF Llandow to become 614's training officer on 3 September 1951. He remembered, 'In the next few weeks we lost three Vampires, fortunately without loss or injury to anyone, and at the subsequent Boards of Inquiry the pilots and the squadron were held to be blameless.'

Two Vampire Mk3s, with Sgt/P2 Jones leading Sgt/P2 James as his No. 2, were returning at 25,000 feet above total cloud cover for a controlled descent before landing at Llandow. Air traffic control at Llandow was in radio contact, 'homing', to guide them in, when the airfield suffered a complete electrical power failure and lost all radio contact with the Vampires.

'Within fifteen minutes full power was restored to Llandow and, in traffic control at Llandow we could hear Jasper Jones and his No. 2 talking to Filton as they "homed" into the airfield, but the ground-to-air Filton transmissions could not be heard at Llandow.'

The 'homing' Tom remembered 'seemed to be going on for ever, when Jasper called that he could see a "White Horse" on a hillside through a gap in the clouds. As our pair were still at 25,000 feet, the "White Horse" they were referring to had to be the one carved into a hillside in Wiltshire, which indicated that they'd missed Filton.'

Sgt James soon radioed that he was out of fuel, and was going to bail out. The twin booms of the Vampire made bailing-out hazardous but, Tom said, 'He was lucky that day. The last R/T call we heard from Jasper, was that he "was in visual contact with an airfield, and would be landing shortly".'

The airfield Jasper was referring to was Greenham Common, at that time under a reconstruction programme as a NATO airfield. 'There were bulldozers, concrete mixers and mountains of gravel-ballast everywhere, and no clear runways available,' Tom explained.

But somehow Jasper managed to land the Vampire without injury to himself, or anyone else, but his aircraft was a write-off. Both pilots were back at their civilian jobs in a couple of days.

However, three weeks later durin the first weekend of October, there was more trouble for Jasper, this time during 'Exercise Pinnacle' at Thorney Island.

Jones was flying the last aircraft to take off during a Wing scramble, but Tom recalled that when he came into land his undercarriage stayed in the locked-up position. 'No matter what Jasper did to try and shake the wheel down, it remained in that position.' With fuel running low, he was advised to belly land on the grass alongside the runway. This he did successfully without injury, but the rough surface caused irreparable damage to the new Vampire FB5 and the aircraft was written off. However, Jasper was soon back on flying duties.[26]

The year 1952 began with the squadron entertaining officers from a visiting Royal Navy vessel, HMS *Ulster*. The adjutant, Flt Lt Peter Latham, who later became an Air Vice-Marshal and was the leader of the Black Arrows formation aerobatic team after leaving 614 Squadron,[27] gave a demonstration in the Meteor to members of the crew.

In a near-miss on 6 April, CO Sqn Ldr McHardy lost both engines in his Meteor T7 at 20,000 feet. The veteran pilot made a wheels-down landing. The diarist wryly added that the CO 'is to enter a Meteor in the National Gliding Contest'.

There was work with the USAF on 'Exercise Ardent' in October. Spread over two weekends, nine Vampires flew over to Filton from where they would be based. The CO started well, with an intercept of a giant ten-engined B-36 bomber. There were several scrambles and interceptions made on B-50s, Anson trainers and Canberra bombers with 'kills' claimed on all types.

The year ended, though, with the lament that, 'with regret, the squadron has to record the departure of three of its pilots during the month, including the boss, Sqn Ldr McHardy. A farewell party was held for the CO who said the squadron should keep its place as one of the best auxiliary units.'

In the new year a new CO arrived, Sqn Ldr JR Spiers, and praise came for Flt Lt Latham's 'considerable skill' in saving a Vampire after engine failure shortly after take-off, when he safely landing at Llandow.[28]

By summer, preparations began for the Royal Review at Odiham for the new Queen's Coronation.[29] Held on 15 July 1953, the Coronation Review of the RAF featured a flypast by some 640 aircraft.[30] This vast parade of aircraft and personnel, including six Vampires from the squadron, saw hundreds of gleaming aircraft lined up for inspection.

The Royal duties continued in July with a flypast over Cardiff for the state visit of HM, the Queen. All pilots 'felt honoured to have this opportunity of showing their formation prowess to the sovereign and it was later learned from the Lord Lt [of Glamorgan] that the Queen had remarked on the high quality of the formation flying.'

Later, there was co-operation with 501 Squadron. Personnel 'worked very hard' in providing the honour guard for the AOC's visit and inspection, which was done as a Wing at Filton. It was admitted that working at Filton required tricky planning, as personnel spent much time travelling to Bristol (at a time before the Severn Bridge was built).

Weather hit the success of the Bruggen camp in July with some terrible accidents later flying with 501. In November 1954, a Vampire from 614 collided with another from 501 at 36,000 feet over the Isle of Wight. P/O Hunter of 614 flew his machine, missing the whole of the starboard boom and most of its tailplane, back to Bristol and after a struggle, bailed out at 10,000 feet. Despite his oxygen bottle getting caught in the cockpit coaming, when he tried to bail when inverted, he eventually fell from the aircraft which crashed in Gloucestershire. The 501 Squadron pilot, P/O Nigel Palmer, was killed.

Preparations for the 1955 annual camp were complex. Taking place at Gibraltar, it would mean ferry flights as long as those made by the squadron during the war.

The airlift to Gibraltar was to begin on 29 July when Vampires and a single Meteor would depart from Llandow for Tangmere. Stopping there overnight, they would fly to Istres in France the next day for refuelling, and then to Gibraltar. The groundcrew would fly in Hastings transports, flying non-stop from Llandow.

This contrasted with 614's perilous transit from the UK to

North Africa over occupied Europe in 1942. It contrasted, too, with 1938, when 614's Hector biplanes had ambled their way to annual camp. In 1955 the jets would fly faster, higher, with more rigid control and far more onerous safety standards.

The squadron settled into a routine of hard work and play. SAC Les Cooper remembered the summer camp fondly, 'The announcement of a trip to Gibraltar was greeted with joy and we all obtained passports to enable us to visit Spain, and Tangiers in Morocco.

'The first town across the border in Spain was La Línea, and the first hostelry was Dick's Bar. It had a fearsome reputation because naval types from all nations spent their time and money there.'

On Gibraltar itself their days had an early start, 05.00, so that they could complete the flying programme by lunchtime. Then they were free for the rest of the day. 'A favourite place on Gibraltar,' he recalled, 'was Catalan Bay for swimming and afterwards a meal in a local café. I can still taste the swordfish steaks.'[31]

The squadron flew a total of 300 hours in Gibraltar, with each pilot flying an average of 20 hours. There were 129 effective shoots with 13,304 rounds fired. These figures show the amount of opportunities available in that concentrated period of their time in Gibraltar, and the value of these annual camps. Not long after the squadron's return from the Rock, they would have a new boss. On 22 August, Sqn Ldr Howe assumed command after Sqn Ldr Spiers was posted to the Air Ministry.

By this time the squadron was flying single-seat Vampire FB5s and FB9s, and had acquired a single-seat Meteor F8, alongside the two-seat Meteor and Vampire.

The uniting of 614 and 501 Squadrons to work as a Wing and provide standing patrols was part of a refocussing of the fighter role of the RAuxAF squadrons and their job in the event of war.[32] But doubts emerged as to the suitability of part-time squadrons in operating sophisticated jet aircraft.

A new policy was announced in the House of Commons by Minister of Defence Harold Macmillan, on 1 December 1954. Both the government and the squadrons had, he said, found it 'unpalatable' that the auxiliaries should be denied the opportunity to fight as formed units. New threats were identified which the auxiliary fighter units would have to deal with. One of these was the latest Soviet bomber, the Tu-4, a reverse-engineered Russian adaptation of the B-29.[33] This was the aircraft that fighters like those of 614 had to intercept and shoot down.

The new year of 1956, however, would see further turbulence in what would be an unsettling year for 614. No flying was carried out in January or February as the squadron had put all its aircraft into storage at RAF St Athan. There were 33 postings away from the squadron and no postings to the unit.

By October, news was received that Llandow would be closing down in March 1957. The writing, perhaps, had been on the wall for some time.

The Baker-Carr Report of 1953 had concluded that, because of the limited amount of flying time available to auxiliary pilots, future aeroplanes would be too complex for them to handle competently. He said that the 20 Auxiliary fighter squadrons should be disbanded. His recommendation was not endorsed at the time, but Baker-Carr's doubts about part-time squadrons became the subject of a separate study for the next four years.[34]

The fears culminated on 15 January 1957 with the announcement that all flying squadrons of the RAuxAF were to disband, and that flying training was to cease. The disbandment was to be effected by 10 March.

An airmen's party was held in Cardiff on Friday, 8 March, to commemorate the closure of the squadron. A dance was held at the officers' Mess in Llandaff with 'all the local dignitaries invited so that a farewell could be said to the many people who worked with and for the squadron.'[35]

The verdict on RAuxAF flying squadrons came in a Defence White Paper that March. In it, Minister for Defence Duncan Sandys set out his vision for the future of the British military. As rockets looked set to replace bombers, it was felt there was no place for first-generation fighter jets like the Vampire and the Meteor that the RAuxAF was flying. The missile-armed, radar-equipped fighters and missiles were seen as too complex for part-time airmen to master. This vision, whether true or not, had sounded the death knell for squadrons like 614.

The gates were closed on Llandow, Number 614 Squadron, and much of the Royal Auxiliary Air Force, ending a story which began in 1937.

CHAPTER 11

Interregnum (1957–2014)

THE GATES OF Llandow may have closed in 1957, but the story didn't end there. An association was formed by ex-members keen to keep in touch. They established a memorial chapel in Llandaff Cathedral and hoped to have a squadron standard awarded, but were refused by the Secretary of State for Air.

Not disheartened, they enlisted the help of south Wales MP, George Thomas, later Lord Tonypandy. In the House of Commons, on 10 May 1961, he asked Mr Willim Taylor, the Under-Secretary of State for Air, whether he'd reconsider and recommend the award of colours for the now-disbanded squadron in view of the fact that it was the only one in Wales.

He replied, 'Under Queen's Regulations, squadron standards are awarded to operational squadrons which have been in existence for 25 years.' As 614 had been existence for only 20, he could not recommend an exception be made, but sympathised with their desire and was 'seeing whether there is any other way in which it can be met'.[1]

The Air Council later agreed that 614 should be allowed the special award of a white RAF ensign in recognition of the squadron's service, having not been in existence long enough to earn a standard. It was presented to the association by Air Officer Wales, Air Cdre M KD Porter, on Sunday, 17

September 1961, and then was laid up, or presented to be housed, in Llandaff Cathedral.[2]

In the following years, the association faded but in 1988, others took up the baton of remembrance. A campaign was launched for a memorial to the squadron at RAF Llandow, and also to the other units which served there.

It was the idea of Philip Davies, of Bridgend, who was there on 6 May 1988 when it was unveiled, watched by ex-614 members who served at Llandow, namely Brian Player, Bob Morgan, former pilot and adjutant AVM Peter Latham and Graham Crocker.[3]

Later that year, the appetite to remember grew. The *South Wales Echo* reported on a meeting in October 1988 to relaunch the 614 Squadron Association; ex-members were welcomed by Forrest Burton, a former 614 Vampire pilot.

Alongside him was Caroline Whitfield, a member of the RAF Auxiliary Defence squadron at RAF St Athan. It was her interest in 614's history that brought about the reformed association. This new association would produce a newsletter where former members could swap stories and memories, trips would be organised to their former homes, like Odiham. But perhaps their greatest achievement was the campaign to have a memorial erected at their former home in Pengam Moors. Secretary of the 614 Squadron Association, Colin Brown, said at the time, 'It was Wales' only Auxiliary squadron, so it is only right that it should be remembered.'[4]

The association president, Wg Cdr Glen Pallot, himself an early pilot at Pengam Moors, spoke at the unveiling on 31 August 1996, of those who first joined the squadron in the 1930s. 'With the enthusiasm of volunteers, learning in most cases new skills, under the instruction of regulars, there was built in less than two pre-war years a spirit of comradeship and pride of achievement in running a fully mobile squadron into the field, which we originals feel was passed on to those who came later through its war-time in this country, North

Africa, Sicily and Italy and to its reformation post-war at Llandow, up to disbandment in 1957. That spirit is carried on by ex-members of our squadron stretching from British Columbia through the UK across to Australia and New Zealand as members of our association.'

The memorial was not the only reminder of 614's place at Pengam that summer. Nearby were the new houses of the Pengam Green estate. After more campaigning by the association, the developers agreed to name the streets after people and aircraft from 614's history. To this day, Pengam Green has a Merrett Court (named after Norman Merrett, the first of the squadron to be killed in the war), Cadman Close (after the first CO), Lysander Court (after its second operational aircraft), Hawker Close (after the makers of its first operational aircraft), Halifax Close (after the heavy bombers it flew in Italy) and de Havilland Road (after the maker of the Vampire jets it flew from RAF Llandow).

The association thrived for a further seven years, forming a vital link between former squadron members living around the world, and bringing together members from 614's different eras. In particular, there were many from the squadron's Pathfinder years who joined alongside the 818s recruited in the 1930s at Cardiff and other theatres. But recognition of 614 Squadron as a Pathfinder unit would be an ongoing struggle for the association. This was symbolically expressed when the RAF Museum at Hendon staged a Pathfinder exhibition featuring squadron crests of the RAF's Pathfinder squadrons, but 614's was missing. The reason given was that it was not part of No. 8 Group (who flew Pathfinder missions from the UK) and therefore could not be considered a Pathfinder squadron. The fact that 614's aircrew undertook training at Newmarket and Upwood under Don Bennett, just as UK-based crews did, and went on to do the same job and were awarded the coveted Pathfinder badge, cut no ice. This lack of recognition produced much anguish in association members – whether they'd served in Italy or not.

Harold 'Barry' Barrington flew as aircrew with 614 in Italy, and was in no doubt that he served in a Pathfinder squadron. He recalled Group Capt (retd) Landon, who was his Flt Cdr at Amendola, expressing surprise at this attitude. He'd met Don Bennett (Pathfinder Force founder and leader) many times, who had 'always agreed that 614 was a PFF squadron and had not the slightest doubt about its recognition as part of the force.'

Another ally later entered the fray. Nick Budd is the nephew of Ted Budd, who was tragically killed when he bailed out of a Liberator as it plunged towards the Adriatic in March 1945. Nick's research uncovered much about the squadron's status, which confirmed that it was indeed recognised as a Pathfinder squadron. Crucially, he discovered there was support from the Chief of the Air Staff ACM Charles Portal for the squadron's application in September 1944 for the award of the Pathfinder badge, and of Pathfinder privileges. This resulted in the first award of Pathfinder badges and certificates to those qualified by AOC 205 Group, Brigadier Jimmy Durrant, in November 1944, at a parade, as described in chapter nine.

Responding to the distress felt by the squadron's veterans at this lack of acknowledgement, Association Secretary Colin Brown wrote to the then Inspector General of the RAuxAF, Air Cdre Sir Hector Monro, appealing for his help in recognising the squadron's role. He wrote, 'I have had much correspondence from members of our association, especially those overseas members in Australia and Canada, saying how upset they are that we cannot achieve official recognition of 614 Squadron's role. At this time in our lives, time is not on our side and the members of our committee feel that if official recognition of 614 Squadron as a Pathfinder squadron is not forthcoming soon, it will be too late. These association members are proud of the fact that they were Pathfinders, and the fact that all we are asking for them is just a small place on the wall is not much.'

Air Cdre Monro duly pursued this, but was met with the answer that while 614 Squadron 'seems to have a valid claim', the Pathfinder display was about 8 Group. While factually correct, this response would do little to assuage the wider hurt felt by 614 Squadron veterans.

Undaunted, the veterans instead got permission to create a Standard for the Squadron Association which they proudly displayed at parades and functions. But as they all grew older, and their numbers became fewer, inevitably, the association folded once more. Their last meeting was on 21 September 2003 and was marked with a farewell service at Pengam Green.[5]

Beyond the association, the memorials, the Standard and the ensign, the squadron lived on in finished metal. At least two of 614's aircraft survive today and are both from the squadron's time at Llandow. They are Spitfire Mk22 PK624 and Vampire F3 VT812. The Vampire is displayed at the RAF Museum, Hendon, while the Spitfire is at Duxford with the Fighter Collection, and is currently under restoration and is not on display.

The white RAF ensign, fraying with age, has been framed and preserved and now hangs at the squadron's headquarters in central Cardiff. The Association Standard now resides at St Michael's and All Angels Church, Michaelston-le-Pit, which adopted 614 Squadron in view of its important connection to the squadron as the resting place of the first of them to fall in war, Norman Merrett.

CHAPTER 12

614 Reformation (2014 to the present)

IT IS 3 March 2022. The Chief of the Air Staff, Air Chief Marshal Sir Mike Wigston, and RAF top brass are gathered at the university in Cathays Park, Cardiff. Above here, three silver Hawker Hectors of 614 flew in 1938, and were captured by a squadron photographer. Air Officer Wales, Air Cdre Adrian Williams, invited the Air Force's main board to meet in Wales for the first time, and taking centre stage alongside him and the chief was the country's only RAF reserve unit, in its home city. It was quite an occasion.

A video was shown where squadron members told of their service. Through their stories they showed what had been achieved since 614 reformed in 2014. While their wartime forebears served on operations like 'Jubilee', the Dieppe Raid, 'Torch', the invasion of North Africa, these current squadron members served on newer, less familiar operations, 'Rescript', 'Shader', 'Azotize' and 'Pitting'.

Much had been accomplished since 2014. On 10 March that year, two biplanes flew low over 614's former Pengam base, home once to those silver Hectors, to mark a special day. The flypast heralded the rebirth of 614 Squadron exactly 57 years to the day after it had disbanded in 1957. The biplanes of the AeroSuperBatics team came courtesy of 614's new Honorary Air Commodore (HAC), Vic Norman. A renowned air display pilot, it was his team, and he flew one of the biplanes.

It was a spectacular moment to mark a hopeful beginning, but the rebirth didn't begin then.

As the squadron's 1957 demise came from that year's Defence White Paper, its rebirth was signalled in its 2010 successor, *The Strategic Defence and Security Review*. In 1957 the paper tried to predict future defence needs in a changing world. Its 2010 counterpart did so too. Both were informed by financial stringency and technological change. The difference for 614 was that the first meant the end, and the second a beginning.

Part of the review was an evaluation of the UK's reserve forces. The result was *Future Reserves 2020*,[1] which said that reserves, including the Royal Auxiliary Air Force, should grow over the next ten years to re-establish a presence in areas where they'd been withdrawn.

In late 2013, four people gathered at the RAF's force development training centre in Crickhowell for an 'away day' of mountain biking and team-building. It was their job to build the reborn 614 Squadron from the ground up. The four included the new OC, the 17th to hold the position, Wg Cdr Phil Sagar MBE. Sagar joined the RAF as an aircraft technician in 1978. Commissioned in 1983, he enjoyed a full career as an administrative officer until 2011, and had recently been appointed as OC 614 Squadron with a 'blank canvas' remit of 'go to Cardiff and set up a new RAuxAF squadron'.

Now a reservist, he was a management trainer in civilian life and often delivered sessions on Tuckman's theory of how organisations evolve. No. 614's rebirth adhered to the principles he set out, namely, 'Forming, Storming, Norming and Performing'. The forming had already begun with *Future Reserves 2020* and the away days at Crickhowell and other places, to gather thoughts and ideas on how to create the reality from this nebulous vision. The challenge they were set was creating a fully operating squadron by 1 April 2015.

Squadrons would re-appear, not as flying units, but as General Service Support (GSS) squadrons, offering many different trades. No. 614 would be in the second tranche of these squadrons to be launched. The first, Nos 611 and 502 were formed in 2013.

Phil Sagar recalled, 'The GSS concept was challenging – but we had to get our heads around it. We only had the squadrons which launched in 2013 as a guide as to the shape of this type of units.' So began the storming.

Some things were apparent early on. 'We wanted to be based in Cardiff city centre, as St Athan was a little isolated. We had to make it easy for people to get to us,' he said, adding, 'Tŷ Llewellyn Reserves Centre, right in the centre of Cardiff, was ideal.'

As well as Wg Cdr Sagar, that small group consisted of the executive officer Sqn Ldr Matt Best; training officer Flt Lt Gavin Walker and Adjutant Sgt Michael 'Freddy' Fredricksen. They went to Pengam to see for themselves the squadron's birthplace and, although almost all the squadron's veterans had now passed away, their relatives were contacted and artefacts brought together. The squadron's rare white RAF ensign which used to hang in Llandaff Cathedral, was retrieved. A case of models representing the aircraft the squadron flew came from Cardiff City Hall.

In those pioneering days, Sagar remembered, 'Some looked to do the process step by step, first infrastructure, then marketing, recruitment then training and employment. We opted to do it simultaneously!'

As well as that small team, there was the squadron's HAC, Air Cdre Vic Norman. Someone with an impressive aviation pedigree, he was an air display pilot with over 2,000 public displays under his belt. Based at Rendcomb, a former RFC airfield near Cirencester, it has many original buildings preserved, including the officers' Mess, complete with gleaming wooden propeller mounted on its wall. His relationship with the RAF goes back many years. 'I've been doing RAF families' day airshows since 1978 and I would mix with RAF personnel and got know them very well. Many of them, like pilots with the Red Arrows, were young then, but many over the years were promoted many times and had reached high rank.'

Following the publication of *Future Reserves 2020*, Vic was invited to the RAF Club in London where he met some of those people. They asked him if he'd be interested in becoming a HAC. 'I was shocked to be honest, but with my knowledge of aviation and my flying business which evolved around media coverage, they said it would be very good to get me on board. I was then approached by Phil Sagar and he was very keen to have some flying activities around the squadron – especially as 614 would no longer fly.' The first of those activities came that March with that flypast over Pengam.

At the same time, adverts appeared online, in newspapers and on Cardiff buses featuring images of Sgt Fredricksen in three guises as a civilian, in RAF 'blues', and camouflage uniform. A sponsorship deal with Glamorgan Cricket Club was struck. More staff were recruited, W/O Alfie Balzano, F/S Gareth Burton and Cpls Sarah Blackford, Dave Goodall, Ellie Williams and Tim Grahame. Tim looked after those squadron artefacts and gleaned more information on the squadron's history before passing the baton of squadron historian to the author in 2016.

In June 2015, for the official opening of 614 Squadron's headquarters the Secretary of State for Wales, David Jones MP, unveiled a plaque of Welsh slate. He was joined by Air Cdre Williams, a representative of 160 Brigade for the Army, and HMS *Cambria* for the Royal Navy Reserve, veterans' relatives and the Australian Air Attache. This recognising the strong links with Australia the squadron inherited from wartime Italy.

By October, the first recruit attested (swore their oath of allegiance) and towards the end of the month the first group of trainees arrived at St Athan, ready to start basic training.

This first phase spread over four weekends. They were kitted out, taught about the RAF, and schooled in drill by F/S Gareth 'Taf' Burton. A miner's son from Blackwood, Taf's RAF journey began in 1985 when he attested as an RAF Regt gunner at the RAF careers' office in Cardiff.

A long and varied career followed during the Cold War, including when he was tasked with guarding nuclear weapons at RAF Molesworth. As a Rapier missile operator, he served in Germany, deploying to the Middle East in the first Gulf War, and later to Kuwait immediately after the 9/11 attacks. He was among the last to leave Basra in 2008 as the UK forces left Iraq, taking with them the memorial wall which is now at the National Memorial Arboretum. During the war in Afghanistan he deployed to Kandahar, working on the 'Hearts and Minds' programmes with locals to build infrastructure. In 2014, he returned to Wales to become the squadron's Flight Sergeant, a post he still holds. Joining a Welsh squadron was a great moment. 'I'm a proud Welshman and it means a lot to me to serve in Wales' only RAF reserve unit and to help instil that pride into our recruits.' That for him is the key, 'Serving in the RAF for me is all about pride, in your service, yourself, your work and your squadron.'

On joining the RAuxAF, reservists commit to 27 days' service per year, passing a fitness test and maintaining their qualifications. Much of this approach is unchanged from the earliest days of what was the Auxiliary Air Force when formed in 1924. The aim is training and preparing people, in their spare time, so they can become a part of the wider RAF when required. Additionally, if those 27 days include 15 days' continuous work or training, they receive a tax-free bounty of

£444 in the first year, rising to £1,759 after five years' service. This compares to £3 in 1924, rising to £5 for years of further service.

The first recruit was Callum Davies and he was joined on that first course by Cardiff native Joe Barber. Now a corporal, he recalled going to St Athan that October. 'It was great to be one of the first, but there were still apprehensions. In the kit list it said you needed to have padlocks for your locker and I had to make a detour on the way to St Athan to get them. Not that they were ever needed!' He remembered the TV programme *Bad Lads Army* and wondered if it would be like that, with 'Corporals shouting in your face'. But fears were soon dispelled. 'We had Wg Cdr Sagar and Sqn Ldr Best taking lessons and we started to learn the basics of being in the RAF.'

But you couldn't get too comfortable, he recalled, and they all had to work hard to get the basics right. 'We struggled with the about-turn on the march and we'd practise in the block, marching up and down the corridor.'

But they passed the course. Joe and all who completed training at St Athan would go to RAF Halton in Buckinghamshire for 15 days. There they'd learn weapons handling, first aid and CBRN (dealing with Chemical, Biological, Radiological and Nuclear attacks). If they completed this, they'd undertake trade training. Recruits come from all walks of life and some serve in roles they already do as civilians while others gain new skillsets. Medical or media specialists, for example, must be qualified in their trade, while those in trades like a driver, HR specialist and RAF police must complete full training. For Joe it meant a year of learning to be an intelligence analyst.

Having completed their training, reservists began to take on tasks supporting RAF operations across Wales, the wider UK or overseas. And since then, those tasks have been many and varied, with personnel deployed to the US, France,

Norway, Ethiopia, South Korea, Egypt, the UAE, Oman and Canada among others.

By now the stage of 'Performing' was being reached and the first 614 deployment overseas since 1957 came in April 2015 when photographer LAC Cathy Sharples went to Estonia as part of 'Operation Azotize' – the RAF contribution to NATO air policing in the Baltic. The author became the first squadron member to deploy to Africa since 1943 when he delivered training in Ethiopia to officers from African Union states, sending peace-keeping forces to Somalia in January 2020.

Both these deployments came via the squadron's Media Flight which recruits journalists, photographers and videographers from civilian life to be trained and deployed to cover RAF operations wherever they may be. It was led from 2015 until 2022 by long-serving ex-regular Sqn Ldr Andy Pawsey. Before joining the reserves, he played a leading role for many years in PR, air shows and events for the RAF.

Recruitment of media specialists by the RAF began in the Second World War when journalists were trained and commissioned as officers so that the RAF could have uniformed reporters on the front line. After the war the RAF maintained a cadre of journalists in uniform. The unit operating them became No. 7644 (VR) Public Relations Squadron, RAuxAF, and 614's Media Flight is part of RAF media reserves which is centred on this squadron.

Wherever reservists serve, the process is similar to that of 1939. Then the squadron was mobilised as a whole but now mobilisations tend to be by individuals released from their job to work full-time for the RAF for up to six months. In 2020, however, 614 deployed a group for the first time since the 1950s.

The fulfilment of Tuckman's metaphor really came that year. As Covid-19 swept the country and the UK went into lockdown that March, the military was asked to shoulder much of the burden – by building field hospitals, solving logistics problems and running Covid testing. This was when the squadron began performing as a whole to support this effort. Phil Sagar recalled, 'We could only deploy people who were fully trained, and the fact that we had enough trained people for two teams ready to mobilise showed just how far the squadron had come,' adding, 'The performance of the deployed teams was phenomenal – under difficult conditions.'

Cathy Sharples, now a Sergeant, led 614's detachment, operating a mobile testing unit in the Midlands. There was trepidation, Cathy remembered. 'We were very much going into the unknown at the time. None of us knew much about Covid, but we all knew we wanted to contribute in whatever way we could.'

One of Cathy's team was Cpl Bradley Fletcher. The intelligence analyst from Devon is one of 614's members who comes from across the Severn as the squadron recruits from the South-West as well as Wales.

He recalled dealing with members of the public in that strange and uncertain time. 'Some were grumpy and some were terrified. At times there were a lot of desperate people showing up and we'd do our best to reassure them.'

Cathy paid tribute to her team, 'Everyone did an exceptional job under really challenging circumstances.'

Living together in a lockdown bubble, Bradley was impressed with the group's camaraderie. 'For those three months, we supported each other and became self-sufficient. We had a real sense of belonging and we depended on each other.'

Cathy tells how the connection they made then still remains strong. 'It's a measure of how much the experience brought us together, that the WhatsApp group we set up at the start of the deployment is still going strong nearly four years later. We still chat on it all the time. It's a very special bond.'

A second team mobilised from 614 was led by experienced

nursing officer Sqn Ldr Jilly Pritchard. Joining the RAuxAF in 1993, between 2003 and 2008, she deployed to Iraq in the second Gulf War, twice to Afghanistan and in 2008 to Sierra Leone. Transferring to 614 Squadron in Cardiff soon after it reformed, she became a much-loved figure, dispensing care and advice. Many in the squadron owe their medical knowledge to her first aid lessons.

Mobilised twice during the pandemic, to NHS Gold Command HQ in London and to Manchester, she relished the challenge. 'It was an amazing opportunity to use my nursing skills and it showed the resilience, flexibility, and adaptability of the reserves. Within a very short time, they were up and running at the testing centres.' In recognition of her distinguished service to the RAuxAF, she was awarded the King's Volunteer Reserve Medal in the King's New Year's Honours List, 2024.

Another who was mobilised was Flt Lt Jude Simpson. Originally an RAF regular physical education officer, she joined 614 in 2014 and is its Adjutant. She first went to support the Army running test centres in Merthyr Tydfil in late 2021. Subsequently, she deployed to the Welsh Government in Cardiff in an RAF team managing military support for the vaccination programme in Wales.

Jude recalled, 'This ran from January through to April and we saw the rollout of the vaccine from Cardiff, all the way through to Llandudno in the north.' She was delighted to be able to support the people of Wales. 'I was extremely proud as a member of 614 Squadron to be able to serve in Wales where I'm based, and to actually help the community.'

Alongside Jude at 160 Brigade HQ in Brecon were Sgt Andrew 'Shaggy' Oldham and Cpl Tom Woodward. An RAF Regt gunner, Tom was training NCO with the squadron until 2020. After joining the RAF in July 2007, he deployed on three tours of Afghanistan between 2008 and 2010 before joining the squadron April 2018. After a long career in the RAF Police from 1988, including deployments to the Falklands and four tours of the Middle East, Sgt Oldham was responsible for employer engagement after joining the squadron in 2016. This vital role brings employer support – key in ensuring reservists can have full careers in the RAF – which means they, in turn, can bring much to their day job. A promotional video for the Reserves that Sgt Oldham set up focused on members of 614 and their employers. One of them was AS1 Michelle Lewis.

A caseworker for Hefin David, a member of the Senedd (Welsh Parliament) for Caerphilly, Michelle mobilised to work on testing units during Covid with Sgt Sharples, and also supported the Welsh Ambulance Trust. 'It's a real privilege to help people – that's why I volunteered during the pandemic. I joined the reserves to do things like this – I felt I could help and if something similar happened, I'd do it again.' In the video, both she and her employer told how being a reservist had made her more confident and resilient.

The men and women of 614 take part in all the major ceremonial and commemorative events where the RAF is represented. Among many events to mark the centenary of the RAF in 2018, one saw a baton relay around the UK to cities and towns and places with links to the RAF. An important connection for 614 was recalled when the baton was carried to Willows High School in Cardiff by former pupil and squadron member, Terence Cadogan. Built on the site of the squadron's Pengam Moors home, it's named after Cardiff aviation pioneer Ernest Willows. Terence told how it was 'an absolute honour' to carry the baton to his old school.

An early and very special occasion to mark RAF100 was actually in November 2017. The Wales v All Blacks game at Cardiff's Principality Stadium hosted a large RAF representation – the central band of the RAF, the Queen's Colour Squadron, and 614 Squadron members, who carried the RAF100 flag on to the pitch.

Two of them were Terence and SAC Toby Grainger. A

Welsh-speaking native of Pontypridd, Toby was an ex-regular with 14 years' experience. An aircraft technician at Halton, Brawdy, St Athan and the Falklands, often on Tornado F3 fighters, he left in 2002, but missed service life.

'I wanted to get back that feeling of family that I had in the RAF, the pride that comes with wearing the uniform, the togetherness, the camaraderie.' He joined as soon as 614 reformed in 2014, this time as an MT driver. It was important to Toby that he served in a Welsh unit. 'That togetherness is part of being Welsh, so being in a Welsh squadron makes it all the more so.' That pride in Welshness comes through in his RAF work. 'I always put on a Max Boyce album when I'm driving the squadron to give the younger ones a taste of Welsh culture.' On a recent visit to the D-Day invasion beaches, thanks to Toby, the squadron's coach resounded to the sound of 'Hymns and Arias' as they drove through Normandy.

A forces life always held an attraction for AS1 Malembe Makawa. Originally from Malawi, his father had military experience there. 'From my dad there was a military discipline that stuck with me.' One of the greatest expressions of military discipline is drill and parading and Malembe was able to show these qualities by taking part in what was then the largest parade a squadron member had joined in recent times. The Platinum Jubilee of Queen Elizabeth II saw around 2,000 military personnel march through London in June 2022. It was a memorable day. 'I remember marching down the Mall, facing Buckingham Palace and realised – "this really is it – this is what we've trained for". We had the order "Eyes right" and there was Prince Charles. It was incredible.' Sadly, his father died in 2019, just before he joined the reserves. My aunties often say, 'If only he could see you now.'

Another great parade came the following year when, after the sad death of Her Majesty, King Charles was crowned in May 2023. Cpl Emily Brown represented the squadron. A learning technologist at Cardiff and Vale College, as a reservist she's a physical training instructor who also mobilised during the pandemic. An accomplished athlete, she represented the RAF in middle- and long-distance running, including at the London Marathon and the Braemar Games in 2018. 'It's really quite special to pull on the jersey and represent the RAF and, as a reservist, I feel very lucky to get these chances,' she said.

For the Coronation, her physical fitness stood her in good stead. Another taking part, AS1 Jonathan Harries, recalled that the training over eight days was tough. 'We marched 138km in total, with the day of the parade being 15.4km.'

The four on the parade, AS1 Harries, Cpls Brown, Toby Grainger and Jon Bridge, were bussed from their base at RAF Halton to London very early on the morning of the Coronation for a spectacular day that will live long in the memory.

Emily recalled, 'Marching down the Mall to Trafalgar Square, the music announcing that the King had been crowned was playing over the PA. It's a moment I'll never forget. The hairs stood up on the back of my neck. I was so proud.' An appropriately crowning moment of the day was the three cheers given by the massed ranks of military lined up in front of Buckingham Palace. 'It was incredible to be in that moment and to look back and think "I was part of that".'

Two of those who paraded for the Coronation served in support of RAF operations overseas. Both Toby and Jon were mobilised to RAF Akrotiri in Cyprus where they supported 'Operation Shader'.

Jon is a former military policeman and another who joined in 2014. He said, 'As an MT driver we get the chance to take on many tasks in different places. I spent four months in Cyprus refuelling RAF aircraft like Typhoons, often on combat operations. It was definitely the most enjoyable part of my career as an MT driver and was a real challenge. I've also served at Al Minhad Air Base in the UAE.'

There were other squadron members who supported 'Operation Shader' at Akrotiri. Cpl Kristen Evans said that it was something she always aimed to do. 'When the chance came, I jumped at it.' The work was often fast-paced. 'I didn't realise at the time but, going back to normal life, I realised what a pressurised situation it was – but that, and the chance to go on operations were reasons why I joined in the first place.'

Getting her campaign medal was memorable. 'My grandad was in the RAF and his medals are at the RAF Halton museum, so to have that tangible proof of what I've done as well is special.'

Intelligence analyst Cpl Gareth Jones enjoyed his time there and was welcomed as part of the squadron he served with. 'Even though I mobilised as an individual and the only reservist – we were all in the same boat together and we bonded really well.'

His time in Cyprus was also fast-paced – one of the busiest periods on 'Operation Shader' – and using his civilian skills in project management, he'd made aircrew briefings sharper and ensured they got information more quickly. After returning to his civilian job, he was surprised to get a call from a Group Captain. 'It was completely out of the blue,' he recalled. 'It was to tell me that I'd been awarded a commendation for my service on the operation. I was over the moon to get it,' he admitted. 'It shows the extra a reservist can bring to the table.'

Another who deployed to Akrotiri is Cpl Dan Horwood. Like Bill Hooper all those years ago at Pengam Moors, his job with machines complements his role in 614 Squadron. He is a senior systems engineer with General Dynamics in Merthyr Tydfil, making armoured vehicles for the Army. 'Being a reservist complements what I do in my day job really well – it's more hands-on with the RAF but still mechanically orientated. Being with 614 helped get me my job. They saw I was willing to put myself in difficult situations and challenge myself.' Squadron members work on RAF operations wherever they're needed. AS1 Steff Morgan, who works in personnel support, was mobilised to Marham in Norfolk, the home of the RAF's latest jet. 'After being a reservist for three years, my civilian employment changed, giving me the chance to do more reserve service. It was fantastic to work on such a busy station and home of the F-35. I love the opportunity that being a reservist gives me.'

In July 2021, Wg Cdr Phil Sagar retired, to be replaced by Wg Cdr Olly Walker. An RAF Regiment officer, his early appointments included being a Flight Commander responsible for nuclear weapon security, and Adjutant of the Queen's Colour Squadron. He was later Chief of Staff on No. 8 RAF Force Protection Wing at RAF Waddington. Promoted to Wing Commander in 2019, he has been deployed to Kuwait, Iraq and Afghanistan.

One of his first tasks, alongside Wg Cdr Sagar and the HAC, Air Cdre Vic Norman, was to present commendations from the Queen's Birthday Honours List to squadron members. They went to all who'd volunteered on 'Operation Rescript' and to two members of the Media Flight, AS1s Gemma Williamson and Anna Harris, who helped produce a virtual Royal International Air Tattoo after the live version was cancelled due to Covid. There were also 'Operation Shader' campaign medals. Also recognised were HQ staff who worked tirelessly to support those mobilised. Led by the Executive Officer, Sqn Ldr Adam Davies, there was Sgt 'Freddy' Fredricksen, who had appeared on those early bus adverts. As well as F/S Burton, there was MT specialist Sgt Will Williams and Cpls Ellie Lewis, Natasha Moore and Sgt Kelly Veryard. She became part of 614 in September 2017, having joined the RAF in April 2008 and was deployed twice to Afghanistan. As well as sharing the team commendation, she later received an individual commendation for her outstanding work in supporting 614 personnel on 'Operation Rescript'.

Cpl Natasha Moore was another to receive a commendation that day. Joining the RAF in June 2001, she became part of 614 in February 2018. Before that, she deployed to Afghanistan in 2008, before going to St Athan, the squadron's parent unit, where she often helped with 614 members on HR matters.

Many in 614 knew personnel there and some of those familiar faces would join 614 as full-time HQ staff. One of them was Cpl Gareth Sutton. As a 19 year old, he left his New Tredegar home to join the RAF Regiment in 1999. He joined the jump-capable No. 2 Squadron and, after a deployment to Kosovo, completed his parachute course. Just over a year later he would be part of the first drop of British paratroopers in armed operations since the Suez Crisis in 1956, when he parachuted into Sierra Leone. Having served in the second Gulf War in Iraq, as part of 16 Air Assault Brigade, he became part of the RAF Regiment team at St Athan before joining 614 Squadron in September 2021. Another welcome arrival from St Athan was HR specialist Cpl Carly Smith who, having joined the RAF in 2009, was based at St Athan from 2017 until joining the squadron in 2021.

On taking the helm of 614, Wg Cdr Walker said, 'I inherited from Wg Cdr Sagar a squadron that was well established – he'd done a superb job. My job was to take 614 to the next phase and ask, "Where are we going next?"

'That's why I set up "Project 614 2030" to create a vision for what 614 Squadron could look like in the coming years. That could mean expanding into new roles, like space operations and increasing recruitment from areas like Swansea, the wider M4 corridor and Bristol. There are fantastic skills bases there with firms like Airbus, Rolls-Royce and the MoD at Abbey Wood which we could engage with more.'

Another change was how RAuxAF squadrons were run. Some functions were centralised and undertaken by Reserve Support Wings (RSW). The first was set up in the summer of 2022 at RAF Leeming in Yorkshire where the Northern RSW took in Nos 611 (West Lancashire), 609 (West Riding) and 607 (County of Durham). Since the summer of 2023, No.

614 Squadron comes under the Southern RSW at RAF Brize Norton, with much training and administration now done there.

But for 614 Squadron, the link with Brize Norton goes beyond this. With the closure of St Athan, from August 2023 the squadron began to be 'parented' by the giant RAF transport base. For Wg Cdr Walker, this marks a profound change and a powerful moment for 614 which will open up many opportunities. 'The fact that we now work for Brize Norton gives us a link to the frontline RAF that we didn't have before. The future is very much how we work with them.'

That connection will take many forms. One is the concept of Agile Combat Employment (ACE). In other words, being able to operate from dispersed areas, increasing the resilience of the home base. Another is Dynamic Force Employment, which means using people out of their traditional roles, especially if they have civilian skills in other areas. For example, it could see chefs being trained to drive tankers.

These two concepts have been driven by the fact that the world in the past five years, and especially since the Russian invasion of Ukraine, is becoming a more uncertain and volatile place. That increasing uncertainty means UK defence in general must be ready to be ever-more flexible in the face of a multiplying number of threats – be they from states like Russia or non-State actors.

He sees the relationship with Brize Norton as rekindling some of the shape of the squadron as it used to be. 'As we work more closely with them as a part of ACE, the whole squadron will be a part of that – and will support dispersed operations from there. This is not how we've worked over the past ten years – with individuals deployed to support the RAF. Before 1957, this is how we used to work, when the whole of 614 was a fighter squadron, and all would support that effort.'

As one era begins, the end of another for the RAF in south Wales gives new purpose to 614's place in its home county.

When the RAF left St Athan in August 2023, 614 Squadron became the main representative of the RAF in south Wales. 'That's why I've been very keen that, as soon as parades returned fully after the pandemic, we were represented at occasions such as the Wales National Remembrance Parade in Cardiff,' he says, adding, 'It was a very proud moment for all of us to take part and it was very special too that Brize Norton-based personnel from south Wales joined us for the parade.' Another effect of this change is that the squadron will have its first annual camp this year – their first since 1956 – at Brize Norton. 'This unity of effort will be a return to how 614 used to work.'

He recognises the need to keep changing. 'We live in a world of constant change and it's my job to drive 614 forward with that in mind. That change will bring many opportunities for us to support the RAF and UK defence and to harness the incredible skills our people have.'

One trade is already building relationships with Brize Norton. The squadron's RAF Police Flt, led by Flt Lt Paul Thomas, has worked closely with the regular RAFP section there. He said, 'We've helped out with their dog sections, for instance, and also at their Brize Fest event, which has helped cement links between us.'

He joined 614 Squadron in July 2020, having been a Royal Navy reservist with HMS *Cambria* and later Cardiff's University Royal Navy unit as a navigation instructor. Alongside his day job at Cardiff University, he's also vice-chair of the Wales military education committee, bringing together the three services' university units which acts as a bridge between the military and Welsh universities and colleges.

It's fitting that this story should end where it began – at Pengam Moors. There, squadron members gathered, dodging rain on a typical Welsh summer's day in July 2022, to learn more about their squadron's past.

Personnel researched talks on episodes of the squadron's

history, given at points on the remnants of the airfield's perimeter track. We spoke of the day the squadron marched from there to war in 1939; of the Malcolm raid in the skies above distant Tunisia, and the gallantry of Bruce Prange as he flew his stricken bomber home from a raid over Romania. We spoke too of Bill Irving – the survivor of that raid in Tunisia who came home to lead 614 in peacetime, only to be killed at its home base in a terrible air crash.

Among those presenting that day was AS1 Jack Pritchard, who found strong connections with Sqn Ldr Irving.

'Researching his story was incredibly poignant. Like me, he's from Cwmbran and the fact that after all he'd gone through, he died in peacetime at the station where he commanded our squadron is powerful enough. But the fact he was leading a trip from a pub – you could almost roll a marble to from my house – the Greenhouse in Llantarnam, made it all the more moving.'

He experienced a powerful event himself a few years ago. 'I became a father in 2021. It gives me an extra motivation to show my daughter, when she's older, some of the things I've done.' The MT driver received a Queen's Commendation for his service during the pandemic, so he's made a good start.

Another taking part was one of the squadron's newest recruits, AS2 Sean Russell. It was one of his first training weekends and he found it a powerful experience. 'Everyone made me feel really welcome,' he recalled. 'It was almost like an initiation into the squadron. I felt I understood what I was joining and was lucky to get an opportunity like that.'

The party walked along the former runway. Now a tree-lined trail, it converges with a series of paths in a clearing, at the centre of which stands the squadron memorial. Erected by the veterans who marched off to war from there and who flew those missions long ago, it stands as a reminder of the squadron's roots and its history.

The day finished with the reading aloud of the Roll of

Honour. The list was passed from hand to hand, with each reading four. Those names of the fallen sounded among the trees as all stood around the stone memorial. From Norman Merrett in 1940 to Edward Campbell in 1953 – some of them would have known Pengam Moors and some, from places across the globe, would never have known it. But they were united in serving in Wales' only Reserve Squadron. And at that moment, they were united too with those serving today.

Endnotes

Chapter 1: Formation (1937–1940)

1 Origin of name Pengam: https://getoutside.ordnancesurvey.co.uk/guides/the-welsh-origins-of-place-names-in-britain/
2 Thursby, RC, *Glamorgan Aviation: Eheda*, The History Press, 2002, pp.38–48.
3 Ibid., p.109.
4 Wing Commander 'Jeff' Jefford, '*AIR FORCE RESERVES 1912 TO MUNICH*', RAF Historical Society Seminar, Reserve Auxiliary Forces, p.11.
5 600squadronassociation.com
6 Thursby, RC, *Glamorgan Aviation: Eheda*, The History Press, 2002, p.148.
7 *Western Mail*, 1 June 1937.
8 Thursby, RC, *Glamorgan Aviation: Eheda*, The History Press, 2002, p.149.
9 *614 Sqn Association Newsletter*, Edgar Salmoni, June 1996.
10 *614 Sqn Association Newsletter*, Bill Drummond, December 1993, p.8.
11 *614 Sqn Association Newsletter*, Hector MacDonald, December 1992, p.8.
12 *614 Sqn Association Newsletter*, Cliff Haines, March 1997.
13 *614 Sqn Association Newsletter*, Tom Coles, July 1997, p.9.
14 *South Wales Echo*, 27 October 1988.
15 Bill Hooper, letter on subject of recruitment to 614 Sqn.
16 *614 Sqn Association Newsletter*, Edgar Salmoni, June 1996.
17 Wilkinson, Frances Louise, *The Territorial Air Force: The RAF's Voluntary Squadrons, 1926–1957*, Pen & Sword, 2020, p.24.
18 White, Ian, 'If You Want Peace, Prepare For War. A History of No. 604 (County of Middlesex) Squadron RAuxAF', in *Peace and in War*, 2005, pp.8–9.
19 *South Wales Echo*, 13 November, 1972.
20 Wilkinson, Frances Louise, *The Territorial Air Force: The RAF's Voluntary Squadrons, 1926–1957*, Pen & Sword, 2020, pp.197–98.
21 *South Wales Echo*, 13 November 1972.
22 614 Sqn Association Archive, Edgar Salmoni.
23 614 Sqn Association Archive, Trevor Stone.
24 *614 Sqn Association Newsletter*, Peter Farr, July 1998, p.18.
25 614 Sqn Association Archive, Trevor Stone.

26 Thetford, Owen, *Aircraft of the Royal Air Force Since 1918*, p.292.
27 *614 Sqn Association Newsletter*, Peter Farr, July 1998, p.18.
28 Thursby, RC, *Glamorgan Aviation: Eheda*, The History Press, 2002, p.150.
29 614 Sqn Association Archive, Trevor Stone.
30 Bill Hooper, letter on subject of first flight.
31 'EMPIRE AIR DAY DISPLAYS (ACCIDENTS)', House of Commons Debates, 28 July 1937, Vol. 326, cc 3095-100, *Hansard*, 28 July 1937, retrieved 4 August 2014.
32 Thursby, RC, *Glamorgan Aviation: Eheda*, The History Press, 2002, p.150.
33 *Western Mail*, 30 May 1938.
34 *614 Sqn Association Newsletter*, Edgar Salmoni, June 1996.
35 *614 Sqn Association Newsletter*, Ron Hann, July 1998, p.13.
36 Thursby, R.C., *Glamorgan Aviation: Eheda*, The History Press, 2002, p.150.
37 *Movement Order*, No. 614 Sqn, 19 June 1938.
38 *South Wales Echo*, 13 November 1972.
39 Faber, David, *Munich: The 1938 Appeasement Crisis*, Pocket Books, 2009, pp.356–57.
40 No. 59 Sqn ORB, https://number59squadron.com/1939_phoney_war.html
41 *Western Mail*, 27 September 1938.
42 *Manchester Guardian*, 1 October 1938.
43 Thursby, RC, *Glamorgan Aviation: Eheda*, The History Press, 2002, p.151.
44 *614 Sqn Association Newsletter*, Peter Farr, July 1998, p.18.
45 *Western Mail*, 22 May 1939.
46 Thursby, RC, *Glamorgan Aviation: Eheda*, The History Press, 2002, p.152.
47 Morgan, Dennis, *Cardiff: A City at War*, 1998, p.5.
48 https://www.bbc.co.uk/history/ww2peopleswar/stories/69/a9002369.shtml
49 *Manchester Guardian*, 22 August 1939.
50 Jennings, W Ivor, 'The Emergency Powers (Defence) (No. 2) Act, 1940', *Modern Law Review* 4 (2), pp.132–36.
51 Keith Sword, 'British Reactions to the Soviet Occupation of Eastern Poland in September 1939', *The Slavonic and East European Review*, Vol. 69, No. 1, January 1991, pp.81–101.
52 https://www.bbc.co.uk/history/ww2peopleswar/stories/69/a9002369.shtml.
53 *614 Sqn Association Newsletter*, Hector MacDonald, December 1992, p.8.
54 *614 Sqn Association Newsletter*, Cliff Haines, March 1997.
55 Morgan, Dennis, *Cardiff: A City at War*, 1998, p.6.
56 *614 Sqn Association Newsletter*, Dai Park, Spring 1995, p.14.
57 AIR-27/2120 – 614 Sqn ORB, October 1939.
58 614 Sqn Association Archive, Trevor Stone.
59 *614 Sqn Association Newsletter*, Cliff Haines, March 1997, p.15.

60 *614 Sqn Association Newsletter,* Hector MacDonald, December 1992, p.8.
61 AIR-27/2120 – 614 Sqn ORB, November 1939.
62 AIR-27/2120 – 614 Sqn ORB, May 1940.
63 Ibid.
64 Hunt, Leslie K, *Twenty-one Squadrons: History of the Royal Auxiliary Air Force, 1925–57,* Garnstone Press, 1972, p.417.

Chapter 2: Scotland (1940–1942)

1 Barclay, Gordon, *If Hitler Comes: Preparing for Invasion: Scotland 1940,* Birlinn Ltd, 2013, p.69.
2 Ibid., p.xviii.
3 TNA WO 166/933 War Office, Home Forces, War Diaries, 28th Infantry Brigade HQ, September 1939–July 1940.
4 TNA WO 166/4186 War Office, Home Forces, War Diaries, 5th Queen's Own Cameron Highlanders, September 1939–December 1941.
5 AIR-25/518 – 22 Group 0RB, July 1940.
6 AIR-27/2120 – 614 Sqn ORB, June 1940.
7 Le Cheminant, Sir Peter, *The Royal Air Force: A Personal Experience,* Ian Allan Publishing, 2001, p.13.
8 AIR-25/518 – 22 Group 0RB, June 1940.
9 Barclay, Gordon, *If Hitler Comes: Preparing for Invasion: Scotland 1940,* Birlinn Ltd, 2013, p.71; TNA WO 166/116 Home Forces, War Diaries, Scottish Command, General, August–December 1940.
10 AIR-27/2120 – 614 Sqn ORB, July 1940.
11 Bill Hooper, letter on subject of preparations for invasion.
12 *614 Sqn Association Newsletter,* JAS Watson, Spring 1995, p.8.
13 AIR-27/2120 – 614 Sqn ORB, July–August 1940.
14 AIR-27/2120 – 614 Sqn ORB, August 1940.
15 https://www.un.org/disarmament/wmd/bio/1925-geneva-protocol/
16 'Notification by the British Government to the German Government on the Outbreak of War Relative to the Geneva Protocol', PRO, WO 193/711; *The Times,* 2 September 1939, p 9; Edward M. Spiers, *Avoiding Chemical Warfare 1939–45,* pp.62–88.
17 Paxman, Jeremy and Harris, Robert, *A Higher Form of Killing: The Secret History of Chemical and Biological Warfare,* Arrow, 1982, p.110.
18 Ibid., p.111.
19 PRO, WO 193 – 732, memo dated 30 June 1940.
20 AIR-25/518 – 22 Group 0RB, June 1940.
21 *Royal Air Force Historical Society Journal,* Conventional Weapons, *THE SCI,* Wing Commander 'Jeff' Jefford, p.162. https://www.rafmuseum.org.uk/documents/Research/RAF-Historical-Society-Journals/Journal_45_Seminar_conventional_weapons.pdf, p.163.
22 AIR-25/518 – 22 Group 0RB, May 1940.
23 Paxman, Jeremy and Harris, Robert, *A Higher Form of Killing: The*

Secret History of Chemical and Biological Warfare, Arrow, 1982, p.111; PRO, WO 193/732. 'Memorandum on the use of gas in the defence of the United Kingdom'.
24 AIR-25/518 – 22 Group 0RB, June 1940.
25 614 Sqn Association Archive, 'Gas-spraying exercise', JAS Watson.
26 BBC People's War, 'From Dunkirk to Burma by Hugh Martin,' https://www.bbc.co.uk/history/ww2peopleswar/stories/35/a6965535.shtml
27 Letter, Ron Hann to Colin Brown, 6 July 1997.
28 Charles Bonner memoirs.
29 Le Cheminant, Sir Peter, *The Royal Air Force: A Personal Experience*, Ian Allan Publishing, 2001, p.14.
30 AIR-27/2120 – 614 Sqn ORB, August 1940.
31 Le Cheminant, Sir Peter, *The Royal Air Force: A Personal Experience*, Ian Allan Publishing, 2001, p.14.
32 AIR-27/2120 – 614 Sqn ORB, August 1940.
33 https://www.theglasgowacademy.org.uk/media/5238/ww2-casualties-tga.pdf
34 https://miltonheritagesociety.co.uk/chapters/milton-in-the-wars/edelweiss-in-milford-on-sea/
35 Halstead, Ivor, *Wings of Victory: A tribute to the RAF*, The Right Book Club, 1941, p.194.
36 AIR-27/2120 – 614 Sqn ORB, August 1940.
37 Terraine, John, *The Right of the Line: The Royal Air Force in the European War, 1939–1945*, Hodder & Stoughton, 1985, p.207.
38 Gilbert, Martin, *Finest Hour: Winston S. Churchill, 1939–41*, Heinemann, 1983, p.123.
39 Barclay, Gordon, *If Hitler Comes: Preparing for Invasion: Scotland, 1940*, Birlinn Ltd, 2013, p.57.
40 Gilbert, Martin, *Finest Hour: Winston S. Churchill, 1939–41*, Heinemann, 1983, p.123.
41 Stansky, Professor Peter, *The First Day of the Blitz: 7-Sep-40*, Yale University Press, 2007, p.136.
42 German Preparations for Invasion in 1940, Parliamentary debate, Volume 430; debated on Monday, 18 November 1946.
43 Barclay, Gordon, *If Hitler Comes: Preparing for Invasion: Scotland, 1940*, Birlinn Ltd, 2013, p.57; TNA WO 166/3443 Home Forces, War Diaries, Royal Engineers, No. 2 General Construction Battalion, June–December 1940.
44 Barclay, Gordon, *If Hitler Comes: Preparing for Invasion: Scotland, 1940*, Birlinn Ltd, 2013, p.57; TNA WO 166/993 Home Forces, War Diaries, 137th Infantry Brigade, HQ, August 1939–March 1940; July–December 1941.
45 Barclay, Gordon, *If Hitler Comes: Preparing for Invasion: Scotland, 1940*, Birlinn Ltd, 2013, p.57; TNA WO 166/1022 Home Forces, War Diaries, 153rd Infantry Brigade: HQ, September–December 1939; August 1940–December 1941.
46 Barclay, Gordon, *If Hitler Comes: Preparing for Invasion: Scotland, 1940*, Birlinn Ltd, 2013, p.57; TNA WO 166/3609 Home Forces,

War Diaries, Royal Engineers, 110 Army Troops Company, July 1940 July – December 1941.
47 AIR-27-2120 – 614 Sqn ORB, September 1940.
48 Le Cheminant, Sir Peter, *The Royal Air Force: A Personal Experience*, Ian Allan Publishing, 2001, p.14.
49 Barclay, Gordon, *If Hitler Comes: Preparing for Invasion: Scotland, 1940*, Birlinn Ltd, 2013, p.57; TNA WO 166/552 Home Forces, War Diaries, 46th Division, General Staff, October 1939–March 1940, July–December 1941.
50 AIR-27/2120 – 614 Sqn ORB, September 1940.
51 AIR-27/2120 – 614 Sqn ORB, October 1940.
52 614 Sqn Association Archive, Donald Mackenzie.
53 Le Cheminant, Sir Peter, *The Royal Air Force: A Personal Experience*, Ian Allan Publishing, 2001, p.14.
54 http://surfcity.kund.dalnet.se/falco_bob.htm
55 AIR-27/2120 – 614 Sqn ORB, October 1940 –January 1941.
56 Charles Bonner memoirs.
57 AIR-27/2120 – 614 Sqn ORB, February 1941.
58 *614 Sqn Association Newsletter*, Ron Hann, June 1996, p.6.
59 https://aviationtrails.wordpress.com/2018/06/03/raf-macmerry-the-darker-side-of-war/
60 614 Sqn Association Newsletter, J Baxter, December 1992, p.16.
61 614 Sqn Association Archive, JAS Watson.
62 *614 Sqn Association Newsletter*, Bill Hooper, July 1995, p.25.
63 *614 Sqn Association Newsletter*, John Richards, July 1998, p.25.
64 *614 Sqn Association Newsletter*, J Baxter, December 1992, p.16.
65 614 Association Archive, Jimmy Redfern.
66 Bill Hooper, letter on the subject of time at Grangemouth.
67 https://aviationtrails.wordpress.com/2018/06/03/raf-macmerry-the-darker-side-of-war/
68 AIR-27/2120 – 614 Sqn ORB, March–December 1941.
69 Gunston, Bill, *Classic World War II Aircraft Cutaways*, Osprey, 1995, p.110.
70 Le Cheminant, Sir Peter, *The Royal Air Force: A Personal Experience*, Ian Allan Publishing, 2001, p.15.
71 Ibid., p.16.
72 AIR-27/2120 – 614 Sqn ORB, January 1942.
73 *614 Sqn Association Newsletter*, Tom Hinchey, Spring 1995, p.15.
74 AIR-27/2120 – 614 Sqn ORB, September 1941–February 1942.
75 Le Cheminant, Sir Peter, *The Royal Air Force: A Personal Experience*, Ian Allan Publishing, 2001, p.17.
76 614 Sqn Association Archive, JAS Watson.
77 *614 Sqn Association Newsletter*, Charles Bonner, November 1996, p.24.
78 AIR-27/2120 – 614 Sqn ORB, September 1941–February 1942.
79 Ibid.
80 Paxman, Jeremy and Harris, Robert, *A Higher Form of Killing: The Secret History of Chemical and Biological Warfare*, Arrow, 1982, pp.114–15.

81 PRO, WO 193/711. Meeting of the Chiefs of Staff Committee, 28 December 1941.

Chapter 3: Raids (1942)

1 Longmate, Norman, *The Bombers: The RAF offensive against Germany 1939–1945*, Hutchinson, 1983.
2 Probert, Henry, *High Commanders of the Royal Air Force*, HMSO, 1991. p.25.
3 Barker, Ralph, *The Thousand Plan*, Airlife, 1992, pp.14–15.
4 Ibid., p.48.
5 Ibid., p.128.
6 Messenger, Charles, *Cologne – The First 1000 Bomber Raid*, Ian Allan, 1982, p.31.
7 Barker, Ralph, *The Thousand Plan*, Airlife, 1992, p.51.
8 AIR-27/2120 – 614 Sqn ORB, May 1942.
9 Barker, Ralph, *The Thousand Plan*, Airlife, 1992, pp.57–60.
10 Messenger, Charles, *Cologne – The First 1000 Bomber Raid*, Ian Allan, 1982, p.40.
11 Barker, Ralph, *The Thousand Plan*, Airlife, 1992, p.60.
12 AIR-27/2120 – 614 Sqn ORB, May 1942.
13 https://www.memorialflightclub.com/bomber-command-first-bomber-raid
14 Bowyer, MJF, *2 Group RAF: A Complete History 1936–1945*, Faber and Faber, 1974 & 1979, p.235.
15 AIR-27/2120 – 614 Sqn ORB, May 1942.
16 Bowyer, MJF, *2 Group RAF: A Complete History 1936–1945*, Faber and Faber, 1974 & 1979, p.235.
17 Le Cheminant, Sir Peter, *The Royal Air Force: a personal experience*, Ian Allan, 2001, p.17.
18 Messenger, Charles, *Cologne – The First 1000 Bomber Raid*, Ian Allan, 1982, p.25.
19 Stapleton, Paddy, *Blow Them a Kiss – a memoir*, Fink Creative, 2018, p.14.
20 AIR-27/2120 – 614 Sqn ORB, May 1942.
21 Bowyer, MJF, *2 Group RAF: A Complete History 1936–1945*, Faber and Faber, 1974 & 1979, p.235
22 AIR-27/2120 – 614 Sqn ORB, May 1942.
23 Messenger, Charles, *Cologne – The First 1000 Bomber Raid*, Ian Allan, 1982, p.24.
24 AIR-27/2120 – 614 Sqn ORB, May 1942.
25 Stapleton, Paddy, *Blow Them a Kiss – a memoir*, Fink Creative, 2018, p.19.
26 Barker, Ralph, *The Thousand Plan*, Airlife, 1992, p.130.
27 https://www.memorialflightclub.com/bomber-command-first-bomber-raid
28 Barker, Ralph,*The Thousand Plan*, Airlife, 1992, pp.48–49.
29 Terraine, John, *The Right of the Line. The Royal Air Force in the*

European War 1939–1945, Hodder & Stoughton, 1985, p.489.
30 AIR-27/2120 – 614 Sqn ORB, June 1942.
31 Terraine, John, *The Right of the Line. The Royal Air Force in the European War 1939–1945*, Hodder & Stoughton, 1985, p.489.
32 Ibid., p.490.
33 AIR-27/2120 – 614 Sqn ORB, June 1942.
34 Terraine, John, *The Right of the Line. The Royal Air Force in the European War 1939–1945*, Hodder & Stoughton, 1985, pp.487–491.
35 Bill Hooper, letter on subject of Dieppe raid.
36 Ibid.
37 AIR-27/2120 – 614 Sqn ORB, July 1942.
38 Bill Hooper, letter on subject of Dieppe raid.
39 AIR-27/2120 – 614 Sqn ORB, August 1942.
40 Ibid.
41 Ibid.
42 Franks, Norman, *The Greatest Air Battle: Dieppe, 19th August 1942*, Grub Street, 1997, pp.28–33.
43 AIR-27/2120 – 614 Sqn ORB, August 1942.
44 Franks, Norman, *The Greatest Air Battle: Dieppe, 19th August 1942*, Grub Street, 1997, pp.26–29.
45 Bowyer, MJF, *2 Group RAF: A Complete History 1936–1945*, Faber and Faber, 1974 & 1979,.p.251.
46 Franks, Norman, *The Greatest Air Battle: Dieppe, 19th August 1942*, Grub Street, 1997, p.33.
47 AIR-27/2120 – 614 Sqn ORB, August 1942.
48 Bowyer, MJF, *2 Group RAF: A Complete History 1936–1945*, Faber and Faber, 1974 & 1979, p.249.
49 Franks, Norman, *The Greatest Air Battle: Dieppe, 19th August 1942*, Grub Street, 1997, pp.42–44.
50 Ibid., pp.37–39.
51 https://www.tracesofwar.com/persons/72403/Scott-John-Edgar.htm
52 Bill Hooper, letter on subject of Dieppe raid.
53 Franks, Norman, *The Greatest Air Battle: Dieppe, 19th August 1942*, Grub Street, 1997, p.29.
54 Stapleton, Paddy, *Blow Them a Kiss – a memoir*, Fink Creative, 2018, p.19.
55 *London Gazette*, 6 October 1942.
56 Bill Hooper, letter on subject of Dieppe raid.
57 Bowyer, MJF, *2 Group RAF: A Complete History 1936–1945*, Faber and Faber, 1974 & 1979, p.251.
58 AIR-27/2120 – 614 Sqn ORB, August 1942.
59 Bowyer, MJF, *2 Group RAF: A Complete History 1936–1945*, Faber and Faber, 1974 & 1979,.p.251.
60 Le Cheminant, Sir Peter, *The Royal Air Force: a personal experience*, Ian Allan, 2001, p.18.
61 *614 Association Newsletter*, Fred Bunting, November 1997, p.8.
62 Le Cheminant, Sir Peter, *The Royal Air Force: a personal experience*, Ian Allan, 2001, p.18.
63 Bill Hooper, letter on subject of Dieppe raid.

64 *London Gazette*, 9 October 1942.
65 Stapleton, Paddy, *Blow Them a Kiss – a memoir*, Fink Creative, 2018, p.19.
66 *London Gazette*, 6 October 1942.
67 Bill Hooper, letter on subject of Dieppe raid.
68 Ian Maclaren logbook.
69 *London Gazette*, 6 October 1942.
70 *614 Association Newsletter*, Dai Park, February 1998, p.19.
71 *London Gazette*, 6 October 1942.
72 Bill Hooper, letter on subject of Dieppe raid.

Chapter 4: North Africa (1942–1943)

1 Le Cheminant, Sir Peter, *The Royal Air Force: a personal experience*, Ian Allan, 2001, p.19.
2 Ibid.
3 Charles Bonner memoirs.
4 AIR 47/33: 'Operation Torch', instructions to O.C. No. 326 Wing.
5 Le Cheminant, Sir Peter, *The Royal Air Force: a personal experience*, Ian Allan, 2001, p21.
6 AIR 27/2120 – 614 Sqn ORB, November 1942.
7 Ibid.
8 Stapleton, Paddy, *Blow Then a Kiss – a memoir*, Fink Creative, 2018, p.21.
9 Letter, Fred Bunting.
10 DC Gunby; Temple, Pelham *Royal Air Force bomber losses in the Middle East and Mediterranean, Vol. 2, 1943–1945*, Air-Britain, 2018.
11 Letter, Bill Hooper.
12 Brian Lane Herder, Darren Tan, *Operation Torch 1942: the invasion of French North Africa*, Odford Osprey Publishing, 2017, p.36.
13 Stapleton, Paddy, *Blow Then a Kiss – a memoir*, Fink Creative, 2018, p.23.
14 Le Cheminant, Sir Peter, *The Royal Air Force: a personal experience*, Ian Allan, 2001, p.22.
15 AIR 27/2120 – 614 Sqn ORB, November 1942.
16 Ibid.
17 Letter, Bill Hooper.
18 Hunt, Leslie, *Twenty-one squadrons: the history of the Royal Auxiliary Air Force, 1925–1957*, Crécy Books, 1971, p.347.
19 Charles Bonner memoirs.
20 LAC Andrew Turnbull diary.
21 Stapleton, Paddy, *Blow Then a Kiss – a memoir*, Fink Creative, 2018, p.23.
22 Wallace, Gordon, *Carrier observer: A Back-Seat Aviator's Story*, Airlife, 1993, p.151.
23 *A History of the Mediterranean Air War, 1940–1945, Volume 3.*
24 Butterworth, Aaron, *With Courage and Faith, The Story of No.18 Sqn RAF*, Air Britain, 1989, p.28.

25 Wallace, Gordon, *Carrier observer: A Back-Seat Aviator's Story*, Airlife, 1993, p.152.
26 Bowyer, Chas, *For Valour*, Grub St Publishers, 1992.
27 Wallace, Gordon, *Carrier observer: A Back-Seat Aviator's Story*, Airlife, 1993, p.151.
28 Butterworth, Aaron, *With Courage and Faith, The Story of No.18 Sqn RAF*, Air Britain, 1989, p.28.
29 Bill Irving, Logbook.
30 *A History of the Mediterranean Air War, 1940–1945, Volume 3.*
31 Butterworth, Aaron, *With Courage and Faith, The Story of No.18 Sqn RAF*, Air Britain, 1989, p.28.
32 *A History of the Mediterranean Air War, 1940–1945, Volume 3.*
33 Le Cheminant, Sir Peter, *The Royal Air Force: a personal experience*, Ian Allan, 2001, p.24.
34 Ibid.
35 *London Gazette*, February 1943, p.815.
36 Le Cheminant, Sir Peter, *The Royal Air Force: a personal experience*, Ian Allan, 2001, p.26.
37 AIR 27/2120 – 614 Sqn ORB, December 1942.
38 Ibid.
39 *614 Association Newsletter*, Charles Bonner, November 1996, p.24.
40 Le Cheminant, Sir Peter, *The Royal Air Force: a personal experience*, Ian Allan, 2001, p.23.
41 AIR 27/2120 – 614 Sqn ORB, Febrary 1943.
42 Le Cheminant, Sir Peter, *The Royal Air Force: a personal experience*, Ian Allan, 2001, p.23.
43 Richardson, Largent, Willard, *RAF Wings Over Florida: Memories of World War II British Air Cadets*, Purdue University Press e-books, 2000.
44 Le Cheminant, Sir Peter, *The Royal Air Force: a personal experience*, Ian Allan, 2001, p.23.
45 Grant, R, 'Up From Kasserine Pass', *Air Force Magazine*, 1 September 2007.
46 AIR 27/2120 – 614 Sqn ORB, February 1943.
47 Ibid.
48 Jefford, Wing Commander CG MBE BA, 'Operation Torch – A Specific Instance of RAF/USAAF Co-Operation', *Royal Air Force Historical Society Journal*, 32, 2004.
49 Ibid.
50 Le Cheminant, Sir Peter, *The Royal Air Force: a personal experience*, Ian Allan, 2001, p.24.
51 AIR 27/2120 – 614 Sqn ORB, February 1943.
52 Stapleton, Paddy, *Blow Then a Kiss – a memoir*, Fink Creative, 2018, p.25.
53 Ian MacLaren DFC citation, *London Gazette*, 18 May 1943.
54 Playfair, Major-General Ian Stanley Ord, *The Mediterranean and Middle East: Volume II*, 1941; *The Germans Come to the Help of their Ally, Official Campaign History V.II*, 1941.

55 Le Cheminant, Sir Peter, *The Royal Air Force: a personal experience*, Ian Allan, 2001, p.30.
56 AIR 49/148 – 326 Wing medical report.
57 Le Cheminant, Sir Peter, *The Royal Air Force: a personal experience*, Ian Allan, 2001, p.31.
58 Jefford, Wing Commander CG MBE BA, 'Operation Torch – A Specific Instance of RAF/USAAF Co-Operation', *Royal Air Force Historical Society Journal*, 32, 2004.
59 'US Army in World War II – Mediterranean Theater of Operations: Northwest Africa: Seizing the Initiative In the West', Chapter XIII in *The Occupation of Algiers*.
60 Playfair, Major-General Ian Stanley Ord, *The Mediterranean and Middle East: Volume II*, 1941; *The Germans Come to the Help of their Ally. Official Campaign History V.II*, 1941, pp.337–342.
61 AIR 27/2120 – 614 Sqn ORB, March 1943.
62 Ibid.
63 Largent, Willard, *RAF Wings Over Florida: Memories of World War II British Air Cadets*, Purdue, 2000.
64 Strawson, Jon, *The Battle for North Africa*, Pen & Sword, 2004, p.209.
65 AIR 27/2120 – 614 Sqn ORB, April 1943.
66 Tomblin, Barbara, *With utmost spirit: Allied naval operations in the Mediterranean, 1942–1945*, The University Press of Kentucky, 2004.
67 AIR 27/2120 – 614 Sqn ORB, April 1943.
68 Stapleton, Paddy, *Blow Then a Kiss – a memoir*, Fink Creative, 2018, p.28.
69 AIR 27/2120 – 614 Sqn ORB, May 1943.
70 Ibid.

Chapter 5: Coastal Patrol (1943–1944)

1 AIR 27/2120 – 614 Sqn ORB, May 1943.
2 Ibid.
3 *614 Association Newsletter*, Rex Thompson, March 1997, p.13.
4 AIR 27/2120 – 614 Sqn ORB, July 1943.
5 *614 Association Newsletter*, Rex Thompson, March 1997, p.13.
6 *614 Association Newsletter*, Walter Matthews, March 1997.
7 AIR 27/2120 – 614 Sqn ORB, August 1943.
8 Ibid.
9 *614 Association Newsletter*, Rex Thompson, March 1997, p.13.
10 AIR 27/2120 – 614 Sqn ORB, September 1943.
11 *614 Association Newsletter*, Rex Thompson, March 1997, p.13.
12 AIR 27/2120 – 614 Sqn ORB, September 1943.
13 *614 Association Newsletter*, Rex Thompson, March 1997, p.13.
14 Ibid., p.17.
15 AIR 27/2120 – 614 Sqn ORB, September 1943.
16 *614 Association Newsletter*, Rex Thompson, March 1997, p.17.
17 AIR 27/2120 – 614 Sqn ORB, September 1943.

18 *614 Association Newsletter*, William Fraser, March 1991, p.3.
19 AIR 27/2120 – 614 Sqn ORB, January 1944.
20 Charles Bonner archive.
21 *614 Association Newsletter*, Jim Redfern, March 1997, p.24.
22 Charles Bonner archive.

Chapter 6: Italy, early days (1944)

1 https://www.airforce.gov.au/sites/default/files/minisite/static/7522/RAAFmuseum/research/units/462sqn.htm
2 Granfield, Alun, *Bombers over sand and snow: 205 Group RAF in World War II*, Pen & Sword, 2011, pp.245–48.
3 *614 Association Newsletter*, Haydn Jones, February 1996, p.18.
4 Bowyer, Chaz and Van Ishoven, Armand, *Halifax and Wellington at War*, Promotional Reprint Company, 1994, p.63.
5 Lake, Jon, *Halifax Squadrons of World War 2: No. 14 (Combat Aircraft)*, Osprey, 1999, p.108.
6 Nijboer, Donald, *German Flak Defences vs Allied Heavy Bombers*, Osprey, 2019, p.32.
7 Lake, Jon, *Halifax Squadrons of World War 2: No. 14 (Combat Aircraft)*, Osprey, p.29.
8 Nijboer, Donald, *German Flak Defences vs Allied Heavy Bombers*, Osprey, 2009, p.32.
9 Lake, Jon, *Halifax Squadrons of World War 2: No. 14 (Combat Aircraft)*, Osprey, 1999, pp.8–10, 70.
10 Ibid., p.10–26.
11 AIR 27/2121- 614 Sqn ORB, March 1944.
12 AIR SOM 466/44.
13 AIR 25/818-834 – 205 Group ORB, March 1944.
14 AIR 27/2121- 614 Sqn ORB, March 1944.
15 Lake, Jon, *Halifax Squadrons of World War 2: No. 14 (Combat Aircraft)*, Osprey, 1999, .p46.
16 Egles, Dudley, *Just one of the many: a navigator's memoirs*, Pentland Press, 1996, p.82.
17 Scotland, Tom, *Voice from the Stars: a Pathfinder's story*, Lynwood, W.A., 1998, p.126.
18 Ibid., p.140.
19 *614 Association Newsletter*, Haydn Jones, February 1996, p.18.
20 *An interesting war*, Arthur Spencer International Bomber Command Centre, pp.30–32.
21 Scotland, Tom, *Voice from the Stars: a Pathfinder's story*, Lynwood, W.A., 1998, p.139.
22 *614 Association Newsletter*, Hugh Fleckney, July 1998, p.8.
23 Egles, Dudley, *Just one of the many: a navigator's memoirs*, Pentland Press, 1996, pp.85–86.
24 Scotland, Tom, *Voice from the Stars: a Pathfinder's story*, Lynwood, W.A., 1998, p.154.
25 Egles, Dudley (1996). Just one of the many: a navigator's memoirs. Edinburgh: Pentland Press. p86

26 *614 Association Newsletter*, Hugh Fleckney, July 1998, p.8.
27 Granfield, Alun, *Bombers over sand and snow: 205 Group RAF in World War II*, Pen & Sword, 2011, p.248.
28 AIR 27/2121- 614 Sqn ORB, March 1944.
29 Overy, Ricahrd, *The Bombers and the Bombed: Allied Air War Over Europe, 1940–1945*.
30 Granfield, Alun, *Bombers over sand and snow: 205 Group RAF in World War II*, Pen & Sword, 2011, p.237.
31 AIR 25/818-834 – 205 Group ORB, March 1944.
32 25/818-834 – 205 Group ORB, March 1944.
33 Hugh Fleckney, memoir letter.
34 AIR 27/2121- 614 Sqn ORB, March 1944.
35 Scotland, Tom, *Voice from the stars: a Pathfinder's story*, Lynwood, W.A., 1998, p.148.
36 AIR 27/2121- 614 Sqn ORB, March 1944.
37 Ibid.
38 *An interesting war*, Arthur Spencer International Bomber Command Centre, pp.30–32.
39 AIR 27/2120-2121- 614 Sqn ORB, March 1944.
40 Scotland, *Tom, Voice from the stars: a Pathfinder's story*, Lynwood, W.A., 1998, p.155.
41 AIR 27/2121- 614 Sqn ORB, April 1944.
42 Ibid.
43 Webster and Frankland, *Strategic Air Offensive*, Vol. 6, HMS0, 1961, Annex I, Radar Aids, Annex IV, Bombs & Bombsights.
44 Macdonald, Patrick, *Through Darkness to Light: The night bomber offensive against Romanian oil, 1944*, Images Publishing, 1990, p.74.
45 Scotland, Tom, *Voice from the stars: a Pathfinder's story*, Lynwood, W.A., 1998, pp.139–40.
46 Ibid., pp.156–57.
47 AIR 27/2121- 614 Sqn ORB, April 1944.
48 *An interesting war*, Arthur Spencer International Bomber Command Centre, pp.30–32.
49 AIR 27/2121- 614 Sqn ORB, April 1944.

Chapter 7: Italy, Oil (1944)

1 TNA IIJ9/4 – The Air Battle of Ploiesti – Fifteenth Air Force, p.1.
2 Ibid., p.71.
3 AIR 27/2121 – 614 Sqn ORB, May 1944.
4 Egles, Dudley, *Just One of the Many: a Navigator's Memoirs*, Pentland Press, 1996, p.90.
5 Ibid., p.91.
6 AIR 25/827 – 205 Group ORB, May 1944.
7 AIR 27/2121 – 614 Sqn ORB, June 1944.
8 Ibid., July 1944.
9 AIR 27/2121 – 614 Sqn ORB, May 1944.

10 Ibid., July 1944.
11 AIR 25/827 – 205 Group ORB, August 1944.
12 Stout, Jay A, *Fortress Ploiesti – The Campaign to Destroy Hitler's Oil*, Casemate Publishers, 2003, p.211.
13 AIR 27/2121 – 614 Sqn ORB, August 1944.
14 Macdonald, Patrick, *Through Darkness to Light: The night bomber offensive against Romanian oil, 1944*, Images Publishing, 1990, pp.226–27.
15 Ibid, pp.229–230.
16 http://www.rafcommands.com/database/wardead/details.php?qnum=5864
17 AIR 25/827 – 205 Group ORB, September 1944.
18 AIR 27/2121 – 614 Sqn ORB, August 1944.
19 *Waterloo Region Record*, 25 March 2013.
20 AIR 25/827 – 205 Group ORB, August 1944.
21 F/O Bruce Prange DFC citation, *London Gazette*.
22 Macdonald, Patrick, *Through Darkness to Light: The night bomber offensive against Romanian oil, 1944*, Images Publishing, 1990, p.207.
23 AIR 25/827 – 205 Group ORB, August 1944.
24 Macdonald, Patrick, *Through Darkness to Light: The night bomber offensive against Romanian oil, 1944*, Images Publishing, 1990, p.240.
25 AIR 25/827 – 205 Group ORB, August 1944.
26 Macdonald, Patrick, *Through Darkness to Light: The night bomber offensive against Romanian oil, 1944*, Images Publishing, 1990, p.240.
27 Ibid., p.372.
28 Ibid., p.207.
29 AIR 27/2120 – 614 Sqn ORB, August 1944.
30 AIR 25/827 – 205 Group ORB, August 1944.
31 AIR 27/2120 – 614 Sqn ORB, August 1944.
32 *614 Association Newsletter*, Geoff Curtis, August 1991, p.5.
33 AIR 27/2120 – 614 Sqn ORB, August 1944.
34 *614 Association Newsletter*, Geoff Curtis, August 1991, p.5.
35 Ibid.
36 AIR 27/2120 – 614 Sqn ORB, August 1944.
37 http://www.rafcommands.com/forum/showthread.php?11334-Halifax-JN912
38 Macdonald, Patrick, *Through Darkness to Light: The night bomber offensive against Romanian oil, 1944*, Images Publishing, 1990, p.247.

Chapter 8: Italy, POWs (1944–1945)

1 *614 Association Newsletter*, Geoff Curtis, August 1991, p.5.
2 Clutton-Brock, Oliver and Crompton Ray, *The Long Road – Airmen Prisoners from Stalag VII (Bankau) to Berlin, June 1944–May 1945*, Grub Street, 2013, p.206.

3 *614 Association Newsletter*, Syd Barlow, July 1999, p.5.
4 Egles, Dudley, *Just one of the many: a navigator's memoirs*, Pentland Press, 1996, p.91.
5 Ibid., pp.91–92.
6 *614 Association Newsletter*, Syd Barlow, July 1999, p.5.
7 Egles, Dudley, *Just one of the many: a navigator's memoirs*, Pentland Press, 1996, pp.90–94.
8 Information and telegram, courtesy of Alison Patriarca.
9 Curtis, Geoff, *Pathfinder force Balkans: one airman's WW2 story of survival*, Woodfield Publishing, 1999, pp.31–35.
10 Langford, Elizabeth (ed.), *Remembering Walter Carrington – Keystone of the Alexander Technique*, Alexandertechniek Centrum, 2006, pp.122–23.
11 Clutton-Brock, Oliver and Crompton Ray, *The Long Road – Airmen Prisoners from Stalag VII (Bankau) to Berlin, June 1944–May 1945*, Grub Street, 2013, p.206.
12 *614 Association Newsletter*, Syd Barlow, July 1999, p.5.
13 Egles, Dudley, *Just one of the many: a navigator's memoirs*, Pentland Press, 1996, pp.95–96.
14 Macdonald, Patrick, *Through darkness to light: the night bomber offensive against Romanian oil, 1944*, Pentland Press, 1990, pp.275–76.
15 *614 Association Newsletter*, Syd Barlow, July 1999, p.5.
16 Curtis, Geoff, *Pathfinder force Balkans: one airman's WW2 story of survival*, Woodfield Publishing, 1999, p.45.
17 Macdonald, Patrick, *Through Darkness to Light: The night bomber offensive against Romanian oil, 1944*, Images Publishing, 1990, p.260.
18 https://www.pegasusarchive.org/pow/SL7/cSL_7_Statement.htm
19 *614 Association Newsletter*, Syd Barlow, July 1999, p.5.
20 Egles, Dudley, *Just one of the many: a navigator's memoirs*, Pentland Press, 1996, p.98.
21 Ibid., pp.99–100.
22 Macdonald, Patrick, *Through Darkness to Light: The night bomber offensive against Romanian oil, 1944*, Images Publishing, 1990, pp.264–67.
23 Ibid., p.268.
24 Ibid., pp.270–71.
25 Egles, Dudley, *Just one of the many: a navigator's memoirs*, Pentland Press, 1996, pp.101–102.
26 Macdonald, Patrick, *Through Darkness to Light: The night bomber offensive against Romanian oil, 1944*, Images Publishing, 1990, p.271.
27 Egles, Dudley, *Just one of the many: a navigator's memoirs*, Pentland Press, 1996, pp.102–104.
28 AIR 27/2121 – 614 Sqn ORB, September 1944.
29 https://www.pegasusarchive.org/pow/cSL_7_History1.htm
30 *614 Association Newsletter*, Syd Barlow, November 1999, pp.16–19.
31 https://www.pegasusarchive.org/pow/cSL_7_History1.htm

32 Ibid.
33 *614 Association Newsletter*, Syd Barlow, November 1999, pp.16–19.
34 https://www.pegasusarchive.org/pow/cSL_7_History1.htm
35 *614 Association Newsletter*, Syd Barlow, November 1999, pp.16–19.
36 https://www.pegasusarchive.org/pow/cSL_7_History1.htm
37 Ibid.
38 *614 Association Newsletter*, Syd Barlow, November 1999, pp.16–19.
39 https://www.pegasusarchive.org/pow/cSL_7_History1.htm
40 http://www.ribera-philosophie.at/pdf/stalag17a.pdf – vgl. NARA: RG389/2151
41 Curtis, Geoff, *Pathfinder force Balkans: one airman's WW2 story of survival*, Woodfield Publishing, 1999, p.70.
42 Meyssignac, Porquoi je ne me suis pas evade S. 191.

Chapter 9: Italy, the road to victory (1944–1945)

1 Pitchfork, Air Cdre Graham, 'Airborne sea-mining operations in World War Two', *Royal Air Force Historical Society Journal*, 45, 2009, p.150.
2 Peter Kaák, David Gunby, *Gardening by Moonlight*.
3 Thompson, Wing Commander HL, *New Zealanders with the Royal Air Force (Volume III)*, Historical Publications Branch, 1959, p.215.
4 Pitchfork, Air Cdre Graham, 'Airborne sea-mining operations in World War Two'. *Royal Air Force Historical Society Journal*, 45, 2009, p.150.
5 AIR 27/2121 – 614 Sqn ORB, September 1944.
6 AIR 25/827 – 205 Group ORB, August 1944.
7 Thompson, Wing Commander HL, *New Zealanders with the Royal Air Force (Volume III)*, Historical Publications Branch, 1959, p.215.
8 Pitchfork, Air Cdre Graham, 'Airborne sea-mining operations in World War Two', *Royal Air Force Historical Society Journal*, 45, 2009, p.150.
9 Nijboer, Donald, *German Flak Defences vs Allied Heavy Bombers, 1942–45,* Osprey, 2019, p.33.
10 Scotland, Tom, *Voice from the Stars: a Pathfinder's story*, Lynwood, W.A., 1998, pp.220–21.
11 Ibid., pp.216–17.
12 Dorr, Robert F, *B-24 Liberator Units of the Fifteenth Air Force*, Osprey, 2000, p.14.
13 Imperial War Museum, https://www.iwm.org.uk/collections/item/object/1060036068
14 *614 Association Newsletter*, Bill Scott, Spring 1995, p.16.
15 AIR 27/2121 – 614 Sqn ORB, October 1944.
16 Ibid., December 1944.
17 Brookes, Andrew J, *Air War Over Italy, 1943–1945*, I Allan, 2000, p.133.
18 AIR 27/2121 – 614 Sqn ORB, December 1944.
19 Scotland, Tom, *Voice from the Stars: a Pathfinder's story*, Lynwood, W.A., 1998, p.225.

20 AIR 27/2121 – 614 Sqn ORB, December 1944.
21 AIR SOM 2147/44.
22 AIR 27/2121 – 614 Sqn ORB, December 1944 – January 1945.
23 Scotland, Tom, *Voice from the Stars: a Pathfinder's story*, Lynwood, W.A., 1998, pp.241–42.
24 Ibid., pp.245–47.
25 Williams, Dennis, *Stirlings in Action with the Airborne Forces: Air Support to Special Forces and the SAS During WWII*, Pen & Sword, 2008, p.27.
26 https://www.thinkdefence.co.uk/2014/12/military-pallets-boxes-containers-part-7-air-despatch/
27 *614 Association Newsletter*, Les Sims, July 1997, p.12.
28 AIR-27/2121 – 614 Sqn ORB, October 1944.
29 Ibid., December 1944.
30 *614 Association Newsletter*, Les Sims, July 1997, p.12.
31 Scotland, Tom, *Voice from the Stars: A Pathfinder's Story*, Lynwood, W.A., 1998, p.222.
32 AIR-27/2121 – 614 Sqn ORB, July1944.
33 Spencer, Arthur, *An interesting war*, International Bomber Command Centre, pp.30–32.
34 AIR-27/2121 – 614 Sqn ORB, July 1944.
35 Scotland, Tom, *Voice from the Stars: A Pathfinder's Story*, Lynwood, W.A., 1998, p.222.
36 http://www.rafcommands.com/database/wardead/details.php?qnum=5818
37 http://www.rafcommands.com/database/awards/details.php?qname=WEARING&qnum=178330
38 AIR-27/2121 – 614 Sqn ORB, October 1944.
39 *614 Association Newsletter*, Frank Pepper, December 1992, p.15.
40 AIR-27/2121 – 614 Sqn ORB, September 1944.
41 Ibid.
42 Mikulan, K.; Pavlovi , Darko; Thomas, Nigel, *Axis Forces in Yugoslavia 1941–45*, Osprey, 1995, p.33.
43 Granfield, Alun, *Bombers Over Sand and Snow: 205 Group RAF in World War II*, Pen & Sword, 2011, p.301.
44 *614 Association Newsletter*, Les Sims, July 1997, p.12.
45 AIR-27/2121 – 614 Sqn ORB, December 1944.
46 *614 Association Newsletter*, Les Sims, July 1997, p.12.
47 AIR 25/827 – 205 Group ORB, December 1944.
48 Brookes, Andrew J, *Air War Over Italy, 1943–1945*, I Allan, 2000, p.137.
49 AIR 27/2121 – 614 Sqn ORB, February 1945.
50 Scotland, Tom, *Voice from the Stars: a Pathfinder's story*, Lynwood, W.A., 1998, pp.247–50.
51 AIR 27/2121 – 614 Sqn ORB, February 1945.
52 Scotland, Tom, *Voice from the Stars: a Pathfinder's story*, Lynwood, W.A., 1998, pp.247–50.
53 Ibid.
54 Ibid, p.250.

55 Storr, Alan, *RAAF fatalities in Second World War among RAAF personnel serving on attachment in RAF squadrons and support units*, 2006.
56 http://www.rafcommands.com/database/wardead/details.php?qnum=33210
57 AIR 27/2121 – 614 Sqn ORB, March 1945.
58 Scotland, Tom, *Voice from the Stars: a Pathfinder's story*, Lynwood, W.A., 1998, pp.260–61.
59 Letter to mother of Sgt D Nicholl from 614 Sqn.
60 *614 Association Newsletter*, Bill Scott, Spring 1995, p.16.
61 Granfield, Alun, *Bombers Over Sand and Snow: 205 Group RAF in World War II*, Pen & Sword, 2011, p.317.
62 *614 Association Newsletter*, Fred Seal, July 1997, p.22.
63 AIR-27/2121 – 614 Sqn ORB, April 1945.
64 AIR 25/827 – 205 Group ORB, April 1945.
65 AIR-27/2121 – 614 Sqn ORB, April 1945.
66 AIR 25/827 – 205 Group ORB, April 1945.
67 *London Gazette*, 12 August 1945.
68 *614 Association Newsletter*, Les Sims, July 1997, p.12.
69 AIR 25/827 – 205 Group ORB, April 1945.
70 Ibid.
71 AIR-27/2121 – 614 Sqn ORB, May 1945.
72 Granfield, Alun, *Bombers Over Sand and Snow: 205 Group RAF in World War II*, Pen & Sword, 2011, pp.323–24.
73 AIR-27/2121 – 614 Sqn ORB, May 1945.
74 Granfield, Alun, *Bombers Over Sand and Snow: 205 Group RAF in World War II*, Pen & Sword, 2011, pp.323–24.
75 AIR 25/827 – 205 Group ORB, May 1945.
76 https://www.rafmuseum.org.uk/research/archive-exhibitions/freedom-liberty/the-long-trip-home/
77 Granfield, Alun, *Bombers Over Sand and Snow: 205 Group RAF in World War II*, Pen & Sword, 2011, pp324–25.
78 *614 Association Newsletter*, Les Sims, July 1997, p.12.
79 AIR 25/827 – 205 Group ORB, May 1945.
80 AIR-27/2121 – 614 Sqn ORB, June–July 1945.
81 *614 Association Newsletter*, George Harper, Spring 1994, p.4.

Chapter 10: Llandow (1946–1957)

1 Freeman, Sqn Ldr AF, *The Post-War Royal Auxiliary Air Force. Royal Air Force Reserve and Auxiliary Forces*, Royal Air Force Historical Society, 2003, p.96.
2 AIR-27-2686 – 614 Sqn ORB, January 1947.
3 Congdon, Phillip, *Per Ardua ad Astra, The Handbook of the Royal Air Force*, Airlife, 1994, p.38.
4 AIR-27-2686 – 614 Sqn ORB, December 1947.
5 Ibid.
6 https://vintageaviationecho.com/cliff-spink-spitfire/

7 AIR-27-2686 – 614 Sqn ORB, August 1948.
8 *614 Association Newsletter*, Ron Ridout, August 1991.
9 https://www.rafmuseum.org.uk/research/research-enquiries/history-of-aviation-timeline/british-military-aviation/1948-2/
10 Ibid.
11 Freeman, Sqn Ldr AF, *The Post-War Royal Auxiliary Air Force. Royal Air Force Reserve and Auxiliary Forces*, Royal Air Force Historical Society, 2003, p.101.
12 AIR-27-2686 – 614 Sqn ORB, March 1950.
13 *South Wales Argus*, 13 March 1950.
14 https://www.wru.wales/article/remembering-the-llandow-air-disaster/
15 AIR-27-2686 – 614 Sqn ORB, March 1950.
16 Ibid., August 1950.
17 Moulson, Tom, *The Millionaires' Squadron – The Remarkable Story of 601 Squadron and the Flying Sword*, Pen & Sword Aviation, 2014, p.196.
18 https://www.rafmuseum.org.uk/research/research-enquiries/history-of-aviation-timeline/british-military-aviation/1951-2/
19 AIR-27-2686 – 614 Sqn ORB, January 1951.
20 *614 Association Newsletter*, Geoff Watkins, February 1996, p.21.
21 AIR-27-2686 – 614 Sqn ORB, May 1951.
22 https://aviation-safety.net/wikibase/wiki.php?id=153855
23 http://ciapoldiescorner.blogspot.com/2008/12/614-sqn-aircraft.html
24 *614 Association Newsletter*, Barri Hurford Jones, June 1996, p.24.
25 AIR-27-2686 – 614 Sqn ORB, September 1951.
26 *614 Association Newsletter*, Tom Scott, June 1996, p.12.
27 https://www.telegraph.co.uk/obituaries/2016/12/08/air-vice-marshal-peter-latham-leader-black-arrows-obituary
28 AIR-27-2686 – 614 Sqn ORB, January 1953.
29 Ibid., July 1953.
30 https://www.rafmuseum.org.uk/research/research-enquiries/history-of-aviation-timeline/british-military-aviation/1953-2/
31 *614 Association Newsletter*, Les Cooper, April 1993, p.11.
32 AIR-27-2686 – 614 Sqn ORB, September 1955.
33 Freeman, Sqn Ldr AF, *The Post-War Royal Auxiliary Air Force. Royal Air Force Reserve and Auxiliary Forces*, Royal Air Force Historical Society, 2003, p.106.
34 Ibid., p.105.
35 AIR-27-2686 – 614 Sqn ORB, March 1957.

Chapter 11: Interregnum (1957–2014)

1 https://api.parliament.uk/historic-hansard/written-answers/1961/may/10/no-614-squadron-colours
2 Hunt, Leslie K, *Twenty-one Squadrons: History of the Royal Auxiliary Air Force, 1925–57*, Garnstone Press, 1972, p.354.
3 *South Wales Echo*, 6 May 1988.

4 *South Wales Echo*, 9 August 1996.
5 *South Wales Echo*, 10 September 2003.

Chapter 12: Reformation (2014–present)

1 https://www.gov.uk/government/consultations/future-reserves-2020-consultation

Selected Bibliography

AIR-25/518 – 22 Grp ORB.

AIR 27/2120-21 – 614 Sqn ORB.

AIR 25/818-834 – 205 Grp ORB.

AIR 47/33 – Operation 'Torch': instructions to O.C. No. 326 Wing.

AIR 49/276 - AIR 49 – RAF Medical Services reports.

AIR 23/5667-5696 – Protection of merchant convoys and Air/sea rescue 1942–1944.

TNA IIJ9/4 – The Air Battle of Ploesti – Fifteenth Air Force.

London Gazette

Western Mail

South Wales Echo

South Wales Argus

Barclay, Gordon, *If Hitler Comes: Preparing for Invasion: Scotland 1940*, Birlinn, 2013.

Barker, Ralph, *The Thousand Plan – The Story of the First Thousand Bomber Raid on Cologne*, Airlife, 1992.

Bowyer, Chaz and Van Ishoven, Armand, *Halifax and Wellington at War*, Promotional Reprint Company, 1994.

Bowyer, Michael JF, *2 Group RAF – A Complete History 1936–1945*, Faber, 1979.

Herder, Brian Lane, Tan, Darren, *Operation Torch 1942: the invasion of French North Africa*, Osprey, 2017.

Brookes, Andrew J, *Air War Over Italy, 1943–1945*, Ian Allan, 2000.

Butterworth, Aaron, *With Courage and Faith, The Story of No. 18 Sqn RAF*, Air Britain, 1989.

Clutton-Brock, Oliver and Crompton Ray, *The Long Road – Airmen Prisoners from Stalag VII (Bankau) to Berlin June 1944 – May 1945*, Grub Street, 2013.

Coombs, LFE, *The Lion Has Wings: The Race to Prepare the RAF for World War II, 1935–1940*, Airlife, 1997.

Curtis, Geoff, *Pathfinder Force Balkans*, Woodfield Publishing, 1999.

Egles, Dudley C, *Just One of the Many – A Navigator's Memoirs*, Pentland Press, 1996.

Franks, Norman, *The Greatest Air Battle: Dieppe, 19th August 1942*, Grub Street, 1997.

Freeman, Sqn Ldr AF, *The Post-War Royal Auxiliary Air Force. Royal Air Force Reserve and Auxiliary Forces*, RAF Historical Society, 2003.

Gilbert, Martin, *Finest Hour: Winston S. Churchill, 1939–41*, Heinemann, 1983.

Granfield, Alun, *Bombers over sand and snow: 205 Group RAF in World War II*, Pen & Sword Aviation, 2011.

Gunby, David and Temple, Pelham, *Royal Air Force Bomber Losses in the Middle East and Mediterranean, Volume 2, 1943–1945*, Air-Britain, 2018.

Halstead, Ivor, *Wings of Victory. A Tribute to the RAF*, The Right Book Club, 1941.

Harris, Robert and Paxman, Jeremy, *A Higher Form of Killing: The Secret History of Chemical and Biological Warfare*, Arrow, 2002.

Hunt, Leslie, *Twenty-One Squadrons: The History of the Royal Auxiliary Air Force, 1925–1957*, Garnstone Press, 1972.

Jefford, Wing Commander CG, 'Operation Torch – A Specific Instance of RAF/USAAF Co-Operation', *RAF Historical Society Journal*, 32, 2004.

Jefford, Wing Commander CG, 'Reserve Auxiliary Forces – Air Force Reserves 1912 to Munich', RAF Historical Society Seminar, 2003.

Kaššák, Peter and Gunby, David, *Gardening by Moonlight*, Peter Kaššák, 2016.

Lake, Jon, *Halifax Squadrons of World War II – Osprey Combat Aircraft No. 14*, Bloomsbury, 1999.

Langford, Elizabeth (ed.), *Remembering Walter Carrington – Keystone of the Alexander Technique*, Alexandertechniek Centrum, Leuven, 2006.

Le Cheminant, Peter, *The Royal Air Force: A Personal Experience*, Ian Allan, 2001.

Longmate, Norman, *The Bombers: The RAF offensive against Germany 1939–1945*, Hutchinson, 1983.

Macdonald, Patrick, *Through darkness to light: The night bomber offensive against Romanian oil, 1944*, Pentland Press, 1990.

Messenger, Charles, *Crucial Battles of World War 2: Cologne – the first 1,000 Bomber Raid*, Ian Allan, 1982.

Moyes, Philip JR, *Bomber Squadrons of the RAF and their Aircraft*, Macdonald, 1976.

Nijboer, Donald, *German Flak Defences vs Allied Heavy Bombers*, Osprey, 2019.

Pitchfork, Air Cdre Graham, 'Airborne sea-mining operations in World War Two', *RAF Historical Society Journal*, 45, 2009.

Probert, Henry, *High Commanders of the Royal Air Force*, HMSO, 1991.

Largent, Will, *RAF Wings Over Florida: Memories of World War II British Air Cadets*, Purdue University Press, 2000.

Scotland, Tom, *Voice from the Stars: A Pathfinder's story*, Lynwood, W.A, 1993.

Shores, Christopher et al., *A History of the Mediterranean Air War, 1940–1945, Volume 3*, Grub Street, 2016.

Stansky, Professor Peter, *The First Day of the Blitz: 7 September 1940*, Yale University Press, 2007.

Stapleton, Paddy, *Blow Then a Kiss – a memoir*, Fink Creative, 2018.

Stout, Jay A, *Fortress Ploesti – The Campaign to Destroy Hitler's Oil*, Casemate Publishers, 2003.

Terraine, John, *The Right of the Line, The Royal Air Force in the European War 1939–1945*, Hodder & Stoughton, 1985.

Thetford, Owen, *Aircraft of the Royal Air Force, 1918–58*, Putnam, 1958.

Thursby, RC, *Glamorgan Aviation: Eheda*, The History Press, 2002.

Wallace, Gordon, *Carrier Observer: A Back-Seat Aviator's Story*, Airlife, 1993.

White, Ian, *'If You Want Peace, Prepare For War', A History of No. 604 (County of Middlesex) Squadron RAuxAF, in Peace and in War*, 04 Squadron Association, 2005.

Wilkinson, Louise, *The Territorial Air Force. The RAF's Voluntary Squadrons, 1926–1957*, Air World, Pen & Sword, 2020.

Zaloga, Steven J, *Kasserine Pass 1943: Rommel's Last Victory*, Osprey, 2005.

Index

13 Squadron 45, 88, 113–14, 124, 132, 137
18 Squadron 108–9, 113, 116, 118, 120, 127, 144
34 Squadron SAAF 234
40 Squadron 204, 217
43 Squadron 61
45 Squadron 45
70 Squadron 238, 260
88 Squadron 88, 93, 100
114 Squadron 75, 78, 81, 129
150 Squadron 194
2 Group 47, 53, 75
205 Group 158, 162, 166, 169, 170, 192, 198–9, 204, 217, 233, 235, 242, 244, 246–9, 252, 259–63, 290
225 Squadron 42
462 Squadron 158–60, 162, 166–7, 171
160 Brigade 295, 300
240 Wing 169, 238
326 Wing 104, 108, 113, 117, 126, 129, 137
500 Squadron (County of Kent) 19, 137
501 Squadron (County of Gloucester) 19, 25, 274, 283–4
502 Squadron (Ulster) 19
613 Squadron (City of Manchester) 19, 45
7644 (Public Relations) Squadron 298

A
Abercrave 270
AeroSuperBatics 292
Afghanistan 296, 300, 305–6
Ager, Sgt Denis 78, 89, 111, 135
Afrika Korps 102, 125
Aiming Point 118, 179, 255
Aine Bieda 118
Aircobra, Bell 147
Air Sea Rescue 155
Albania 238
Aldeia Nova 106
Alessandria 182
Algeria 7, 13, 101–2, 104–5, 108, 121–2, 141, 153, 166–7
Algiers 105, 108, 110, 119, 137, 153, 236
Altair 247, 254, 258
Amendola 191–3, 225, 235, 237, 245, 249, 252, 258, 290
Amiens 44
Amsterdam 79, 82
Anderson, Sgt John Hume 204
Andover 69, 92
Angel Hotel 21
Anzio 169
Ardennes 44
Ariana 135
Armadillo 50
Arnim, General 122, 129
Athens 246–7, 249
Attlee, Clement 279
Atwell, P/O Gerald Joseph 168, 181–2

Aubrey-Rees, P/O 108, 139
Avro Anson 29, 282

B
Baelz, P/O 77, 84
Balzano, W/O Alfie 295
Bankau 218, 226
Banks, Leslie 62
Barber, Cpl Joe 297
Bari 224–5, 248, 262, 265
Barlow, F/S Syd 207–8, 212, 218–19, 221, 226–7, 231–2
Barrington, Harold 'Barry' 290
Barry 260, 270
Basra 296
Bastia 269
Bath, W/O Harry 196
Battle, Fairey 29, 36
Battle of Britain 46, 60, 77, 153, 177, 267, 271, 273–4, 277
Baxter, AC2 John 64–5
Bay of Biscay 105–7, 166–7
Beevor, Sgt Paddy 210, 189, 223
Belgium 44, 75
Bennett, F/S Roy 206
Bennett, Wg Cdr Don 161, 163–4, 166, 276
Berchtesgaden 240
Berlin 157, 129, 204, 241, 253
Berwick-upon-Tweed 48
Best, Sqn Ldr Matt 294
Bile a 245
Bioce 252
Bisley, Bristol 7, 102, 104–6, 108–9, 111–120, 123–4, 126–34, 137–41, 143–4, 147–150
  BA731 119
  BA800 113
  BA809 105
  BA825 113–14
  BA869 113
Bizerta 109, 111–13, 117, 134–5
Blackford, Cpl Sarah 295
Blackpool 266
Blackwood 296
Blenheim MkIV, Bristol 29, 54, 67–71, 73, 75–8, 80–4, 87–92, 94–5, 97–9, 101–3, 106, 123, 141, 147
  R3758 77
  V5752 77
  V6002 87
  Z6032 73
  Z6033 70
Blida 104–5, 108–9, 111–12
Blind Illuminator 78
Blohm and Voss 140
Bofors 60, 62
Bologna 253, 261
Bomber Command 74, 76, 81, 159, 160, 163, 242, 268
Bonn 75, 81
Bonner, Charles 72, 104, 109, 111, 117, 119, 128, 130, 134–5, 142, 150, 153, 155, 156–7
Borizzo 142–4, 149–50, 153, 155–8
Boston 87, 92–3, 96, 100, 120, 126, 128–9
Boston, P/O Charles Henry 262
Boulton-Paul Defiant 61
Boyce, Sgt Samuel John 172
Bradford, W/O Mervyn 214
Brain, Sgt 30
Brecon 269, 300
Bremen 78, 82–3
Brenner Pass 237
Bridge, Cpl Jon 303
Briggs, Sgt K 206
Bristol Mercury 37, 67, 90
Brod 190–1
Bronzolo 237–8
Brown, Colin 288, 290
Brown, Cpl Emily 302, 303

Brown, F/S Albert Sidney Harrison 190
Browning 37, 49, 68, 125, 271
Brownless, Flt Lt Sid 199–200
Bruce, Flt Lt Ian Henderson 249
Brunton, F/S Peter 196
Bucharest 189, 192, 195–6, 198, 202, 204, 210, 222, 224–5
Budapest 190, 202, 212–13, 218, 221
Budd, Nick 290
Budd, F/S Ted 255–8, 265
Bulgaria 170–2, 174, 180, 188, 207, 223
Bunting, Sgt Fred 94, 103, 106
Burton, F/S Gareth 295–6, 305
Burton Forrest 288

C

Cadman, F/O Charles 21, 43
Cadman, Sqn Ldr/Air Cdre Richard 21, 24–5, 32, 42–3, 103, 157, 268, 270, 289
Cadogan Terence 301
Cagliari 143
Caldwell-Wearne, F/S Claude 194–6
Calvert, F/O Douglas 217, 219, 222–3
Campbell, Pilot 2 Sgt Eddie 280, 310
Campbell, W/O1 George 112, 117
Campina 189
Canberra 282
Canrobert 7, 112, 117–19, 122
Cantacuzino, Captain Carl 224
Canton 7, 23
Cap Bon 135
Capodichino 262
Cardiff 267–8, 274, 278–80, 283, 285, 289, 291–7, 300–2, 308
Cardiff Municipal Airport 17, 20, 36
Cardiff University 308
Caro, Lt John Leonard 204
Carrington, Flt Lt Walter 214
Cartwright, F/S Ron 215–16
Casarano 238–9
Casarsa 258
Catania 167, 173
Cater, F/S Samuel 190
Cathays Park 29, 274, 292
CBRN 297
Celone 158, 161, 167–9, 172–4, 177, 186, 191, 237
Chamberlain, Neville 33, 37–8, 40
Cheylus 134
Chouigui 113, 116–17
Churchill, Winston 52, 74, 83, 99, 170
Clarke, F/S John William 196
Cleak, Sub Lt Francis 120
Coles, Tom 22
Collins, F/S RB 196
Cologne 75–7, 81
Constantine 120
Cooper, SAC Les 284
Corbi Mari 195
Cormie, Sgt Gordon 204
Coronation 282, 303
Corsica 148
Cote, Sgt RL 196
Cremona 256–7
Crete 266
Crocker, Graham 272, 272, 288
Cromwell 58–9
Curtis, Sgt Geoff 202–3, 205, 211, 214, 218, 230, 232
Cwmbran 7, 114, 268, 275–6, 309
Czech 32
Czechoslovakia 33–4, 38, 191

D

Danube 196, 202, 207, 218, 233–4
Danzig 34, 40
David, Hefin 301
Davies, Callum 297
Davies, Philip 288
DC3 Dakota 142
De Bloeme, Flt Lt PGG 191
de Wesselow, P/O Peter 94, 106, 109
Dear, F/S Norrie 171, 189
Deri, Bargoed 280
Dickson, F/S Vivian 171
Dingwell, F/S RJC 172
Distinguished Flying Cross 101, 103, 120, 125, 140, 198, 205, 242, 260, 264, 267–8, 276–7
Distinguished Flying Medal 206, 236
Distinguished Service Order 261, 277
Dobson, Megan 66
Dolman, Sgt Percy Charles 249
Dorrington, F/O Ronald Kingston 262
Dowding, Air Chief Marshal Hugh 58
Drake, P/O Albert 104
Drax, Admiral Sir Reginald 47
Drummond, Bill 21, 42
Dulag Luft 218
Dumfries Place, Cardiff 20, 22, 268
Duncansby Head 48
Dunkirk 45–6, 52, 114, 276
Dunn, Sgt Harold 249
Durrant, Brigadier 242, 290

E

Earl, P/O James J 249
East Grinstead 99
Eayrs, Wg Cdr 60
Edinburgh 19, 46, 59, 61–2, 64, 67
Edwards, F/Lt Martin 25, 27, 48
Edwards, Flt Lt Albert George 204
Egles, Dudley 163–4, 167–8, 175, 189, 190, 195, 207–8, 210, 217–18, 222, 224–5
Egypt 102, 158, 163, 265, 298
Eighth Army 130, 248, 251, 259–62
Eisenhower, General 121, 126, 133
El Alamein 102, 120
El Guettar 130, 132
El-Adem 158–9, 166
Elizabeth, HRH Princess 274
Elizabeth, HRH Queen 302
Ellwood, Sgt PJ 56
Elvas 106
Empire Air Day 28–9, 34, 36
Enfidaville 119, 132–4
England, F/O Richard 'Dickie' 43, 63, 71, 100
Essen 81–3
Esztergom 234
Ethiopia 298
Evans, P/O Howard Hughes 270
Evans, Cpl Kristen 203
Evans, F/S John Graham 203
Exercise:
  Ardent 282
  Bulldog 274
  Dagger 272
  Emporer 278
  Foil 274
  Pinnacle 281

F

Farr, Peter 25–6, 34–5
Feuersbrunn 248
Fighter Command 18, 58, 61, 273
Filton 19, 274, 281–3
Findlay, P/O John 72–3

First Army 129–30, 132
Firth of Forth 59, 63, 70
Fiume 191, 254
Fleckney, Hugh 166, 169, 172, 174–5, 177, 200–1, 241, 249–51, 264
Fleets Colliery 64–5
Fletcher, Cpl Bradley 299
Flintshire 28
Fochville 133
Foggia Plain 158
Forshaw, Sgt J 198
Foster, F/S Norman 195–6, 218
Foum de Gueiss 116
Fox, W/O Allan 197–9, 126, 129
France 33–4, 38, 40, 42–5, 52, 58, 67, 75, 83, 85, 90, 107, 196, 208, 247, 283, 297
Francis, Flt Lt RE 186
Francis, P/O RD 242
Fraser, Willy 153
Fredricksen, Sgt Michael 'Freddy' 294–5, 305
Freilassing 261
Friston 67
FW190 90, 92, 198

G
Gabes 132
Gafsa 121, 124, 130
Gardening 234
Gas Spray 43, 53–4, 73
Gee 172, 175
Geneva Protocol 51
Genoa 170, 182
Georges, P/O Cyril 82, 92–3, 113, 116–17
Germany 18, 32, 34, 37–8
Gifkins, Sgt 79, 89, 91
Giras, P/O George 73
Gittings, W/O Eric 190
Giurgiu 190
Gladstone, F/O Douglas Stuart James 256
Glatton 266
Glover, Robert 153
Goldberg 230
Goodall, Cpl Dave 295
Göring, Herman 57
Gothic Line 248, 251
Gower 277
Grahame, Cpl Tim 295
Grainger, Cpl Toby 301, 303
Grangemouth 46, 48, 50–1, 54
Grant, Ron 29
Graz 262
Greece 245, 249, 251
Greene, W/O Richard A 219
Greenhouse Inn 275, 309
Grierson, Jock 153–4
Griffithstown 276
Grumm, Karl 219
Guernsey 8, 45, 92
Gustav Line 168
Gyor 202

H
H2S radar 159, 162, 164–5, 167, 175–6
Haines, Cliff 22, 39, 42
Halifax, Handley Page 137, 158–62, 164–5, 168–75, 178, 181–5, 189–200, 202–3, 205–7, 214, 234, 236–8, 248–50, 252–3, 289
BB382 172
BB385 171
BB415 172
BB446 172
JN890 171
JN912 203
JN942 191
JN943 181
JP107 190
JP109 214

JP110 194–6
JP111 189
JP135 193
JP183 249
JP223 202
JP225 196
JP227 199
JP232 200, 249
JP234 249
JP259 192
JP275 202
JP279 193, 234
JP282 194–6
JP287 248
JP321 194, 206
JP323 206, 218
JP352 193
Hamburg 76, 204
Hampden, Handley-Page 61
Hanbury, P/O 87
Hann, Ron 30, 54, 64
Hannah, P/O 82, 89, 94, 97
Harper, Al 255
Harper, F/O James Ferguson 56
Harper, George 265
Harries, AS1 Jonathan Johnny 303
Harris, ACM Arthur 74–7, 81–3, 163, 190, 263
Harris, AS1 Anna 305
Harrison, F/O Maurice George 249
Harrow, Handley-Page 84
Hart, Hawker 25–7, 29, 31
Harvard, North American 267–9, 272–3, 276, 279–80
Harvey, F/S BT 144–5
Hawarden 61
Hawick 82
Hawken, LAC Henry George 124
Hayward, Sqn Ldr DG 159
Hearn, Sgt William Frederick 172
Heath, F/S Douglas James 203
Hector, Hawker 26–7, 29–31, 33, 36–7, 39, 41–2, 45, 284, 292
K9784 27
K8111 41
K9740 41
Hednesford 42
Hendon 25, 289, 291
Heyford, Handley-Page 22, 29
Hibbins, F/S Leonard Henry 181
High Wycombe 76
Himmelbett 192
Hind, Hawker 23, 25–6, 29, 31, 41
L7238 41
Hinchey, Tom 70
Hispano-Suiza (cannon) 49–50, 57, 59, 271, 277
Hitchen, F/S Bob 208, 210
Hitler, Adolf 32–4, 38, 40, 42, 52, 57, 129, 219, 240–1
HMS *Cambria* 295, 308
Hodges, F/S Norman Ernest 204
Hodson, Sgt Norman Harold 249
Hodson, Pilot 2 269–70
Holloway, Ted 113–16, 144–6, 150–3
Homes, F/O Joseph Edward Albert 204
Hooper, Bill 7, 23, 28, 49, 65–7, 84–7, 89–91, 96, 99, 101, 107, 110–11, 157
Hopper, Derek ‘Hoppy’ 164
Hornchurch 61
Hornet 273–4
Horsham Saint Faith 67, 272
Horwood, Cpl Dan 304
Howe, Sqn Ldr 284
Huggett, Sgt William 73
Hudson, Lockheed 137–8
Hungary 182, 187, 202, 205–6, 213–14, 218, 225, 249
Hurford Jones, Barri 280

Hurricane, Hawker 18, 27, 46, 51, 56, 61, 97, 165, 246, 278

I

Innerhalpbach 248
Inverness 48, 70–1, 100
Iraq 296, 300, 305–6
Ireland 8, 275
Isaac, LAC H 204
Isle of Wight 283

J

Jackson, F/S Donald Herbert 171
Jagdgruppe 90
James, Sgt/P2 280–1
Jamison, Sgt DH 198
Jarman, Grp Capt 247
P/O Johnson 60
Johnson, Sgt 79, 89, 91, 98
Jones, Cpl Gareth 304
Jones, Haydyn 159, 165, 175
Jones, 'Jasper' 279, 281–2
Jones, F/S Herbert George 195
Jones, F/S HG 196
Jones, F/S John Joseph 181
Jones, Sgt/P2 276
Joseph Sharp, F/S Edmund 249
Juvincourt 75
Ju88 62, 75, 82, 90, 206

K

Kairouan 119, 130–1
Kaisersteinbruch 219, 230–1
Kalamaki 246
Kandahar 296
Kasserine Pass 121–6
Kendall, F/S Frederick 256
Kesselring, Field Marshal Albert 169, 259
Kestrel 23, 26
King's Own Scottish Borderers 89
Kinnaird Head 48
Klagenfurt 248
Kolacin 252–3
Krems Gneixendorf 218
Kursk 177
Kuwait 296, 305
Kwinter, F/O Samuel 203

L

La Marsa 133, 142, 153
Lacey, Flt Lt TK 42, 43
Laird, Grp Cpt JS 60, 191, 213–14, 264
Lake Balaton 202
Langton, Flt Lt Robert D 194–5
Latham Peter 282, 288
Latisana 238
Lawley, F/S Harry Gardner 256
Lawson, F/S Lyle Anzac 171
Le Cheminant, Peter (Jnr) 12
Le Cheminant, Peter 45, 48, 54
Lee, Sgt Walter 171
Leeuwarden 82–3
Leigh-Mallory, Air Marshal 86
Les Andalouses 138
Lewis, AS1 Michelle 301
Lewis, Bob 165, 183–5
Lewis, Cpl Ellie 305
Lewis, F/S David John 172
Liberator, Consolidated 168–71, 194, 199, 234, 236–8, 240–1, 244–7, 252–6, 258, 260–4, 265–6, 290
KG944 246
KG945 245, 258
KG949 245, 247
KG952 256
KG957 245
KH237 246
Libya 129, 158, 161, 188
Lichfield 266
Lightbody, F/S Robert Thomas 172

Limoges, Sgt 92, 113, 115
Ljes 238
Llandaff Cathedral 287–8, 294
Llanharan 65
Llantarnam 275, 309
Llantwit Major 28, 274
Lloyd, Air Vice-Marshal Sir Hugh 137, 142, 156
Lockwood, Wg Cdr EBR 261, 264
London 19, 24, 31, 39, 43, 57, 106, 160, 214, 278, 295, 300, 302–3
Lord, F/S Norman 194–5
Low Countries 44, 58, 208
Luftwaffe 45, 60, 90, 110, 188, 210, 211, 218
Lysaght, F/Lt 'Johnnie' 63, 71
Lysander, Westland 37–8, 41–2, 44–5, 48–9, 51, 53–6, 59–61, 67–8, 289

M

Macchi 200 143
MacCreery, General 261
MacDonald, Hector 21, 39, 42
Mackenzie, SAC Donald 60
Maclaren, Ian 77–80, 83–4, 89–90, 97, 99, 109, 124–5
Macleod, F/S John Malcolm 181
Macmerry 62–9, 73, 83, 85–6, 100, 114
Macmillan, Harold 285
MacNamara, F/O Brian 61, 77–8, 81
Mae West 138, 151, 161, 208
Magee, P/O John Gillespie 268
Magor 41
Maison Blanche 167, 236
Makawa, AS1 Malembe 302
Malan, Sqn Ldr AAN 43
Malcolm, Wg Cdr Hugh 113–16, 118, 120
Manchester Guardian 33
Marchigiano Mountains 183
Mareth Line 130
Marquis of Clydesdale 24
Martin, Flt Lt GA 267
Martyn, P/O Walter Joseph 195
Massicault 134
Matesevo 252–3
Mathews, Walter 140–3, 148–50
McCormack, P/O John 249
McFall, F/O SE 242
McHardy, Sqn Ldr KH 'Sam' 277, 282
Mckay, Sgt Will Craig 181
McKee, W/O Peter 249
Me109 7, 61, 75, 90, 94, 114, 197, 224
Me110 75, 118
Mediterranean Allied Air Forces (MAAF) 156
Mediterranean Allied Strategic Air Force (MASAF) 158, 162, 169
Mepham, Flt Lt Harry Kenneth 256
Merrett, Herbert 56
Merrett, Ray 38
Merrett, Norman 26, 55–6
Merthyr Tydfil 25, 300, 304
Mestre 181–2
Meteor, Gloster 266–7, 279, 282–86
Meyssignac, Jean-Marie Marcel 232
Michaelston-le-Pit 56, 291
Mihai Viteazui Barracks 217
Mitchell, North American 126, 129
Mitchell, Sgt Harry William James 204
Mitchell Pilot 2 269, 289
Mk XIV Bomb Sight 175–6, 180

Moncrieff, Wg Cdr LWJ 250–1, 260
Monro, Air Cdre Sir Hector 290–1
Montgomery, General 121, 125–6, 129–30
Montrose 48, 55–6
Moore, Cpl Natasha 305
Morgan, P/O David Edward 269–70
Morgan, AS1 Steff 304
Morgan P/O 274, 276–7
Morgan-Smith, Flt Lt P Gilbert 21–2, 28
Morocco 102, 166–7, 284
Mosquito 186
Mosquito, de Haviland 261
Mountain Ash 274
Mutton, F/S David Idris 262

N

Napier Dagger 26
Nazi–Soviet Pact 39
Netherlands 75, 82, 274
Neubeck, Flt Lt William 'Bill' Francis 254, 256
New Zealand Corps 130
Newhaven 86, 164
Newstead, Cpl Harry 204
Nicholson, P/O John 206, 216
Nicholl, Sgt David 257–6
Ninian Park 40
No. 8 (Pathfinder Force) Group 163, 242, 289
Norenius, Lt Gunnar 103–4
Norman, Air Commodore Vic 292, 294, 305
Northwest African Coastal Air Force 137, 156
Norway 44, 47, 161, 163, 298
Notts County 40
Novello, Ivor 62

O

O'Connor, F/S Arthur Alexander 204
Oldham, Sgt Andrew 'Shaggy' 300–1
Operation:
  Azotize 292, 298
  Capri 129
  Exodus 264
  Jubilee 85, 88, 292
  Pitting 292
  Pugilist 130
  Rescript 292, 305
  Rutter 83, 85–6
  Shader 292, 303–5
  Strangle 169–70
  Tidal Wave 188
  Torch 102, 104 , 120, 125, 137
Oulmene 101, 109–23, 125, 130–1, 137–8

P

Palestine 265
Pallot, Glen 25, 28–9, 43, 288
Palmer, Len 167
Palmer, P/O Nigel 283
Pantelleria 173
Parramatta 164
Paris 25
Park, Air Vice Marshal Keith 153, 156
Park, Dai 22, 41, 65, 100
Parliament 39, 264, 301
Parsloe, Henry 114
Partisans 211, 239, 243–6, 252–3, 263
Patriarca, F/S MR 'Ron' 203, 212–14, 231
Peirse, ACM Richard 74
Pembroke Dock 36, 277
Penarth 21, 273, 277
Pengam Moors 7, 8, 17–18, 20–3,

26–31, 34–6, 39–43, 55, 63–4, 72, 100, 278, 288–9, 291–2, 295, 301, 304, 308, 310
Penston 63
Pfaffendorff 228
Phoney War 44, 73, 157
Piave 238
Piorun 152
Player, Sgt Brian 279, 288
Ploiesti 188, 191, 195–6, 199, 201, 222
Plovdiv 172, 174, 180
Podgorica 252–3
Pola 254–6
Poland 34, 39–49, 44
Pont du Fahs 119
Pontypridd 274, 276, 302
Popesti 225
Port Talbot 269, 273
Portal, ACM Charles 74, 290
Porter, Air Cdre M KD 287
Porter, LAC Frank Edward 142
Porter, P/O 81–3, 93
Portomaggiore 261
Portreath 105–6, 166
Portugal 105–6
Potter, P/O 43, 82
Prahova 191
Prange, F/O Bruce 196–8, 309
Pratt & Whitney 236
Prausnitz 230
Prestney, F/O 150
Pridham, F/S GF 172
Prime, Sgt Victor Barwick 181
Pritchard, AS1 Jack 309
Pritchard, Sqn Ldr Jilly 300
Pryor, W/O Arthur William 249

Q
Quevatre, P/O Leonard 92, 113, 115

R
Rabat 166–7
RAF:
Abingdon 164
Acklington 279
Greenham Common 281
Halton 53, 297, 302–3
Hurn 166
Llandaff 268, 285
Llandow 6, 8, 267–71, 274–6, 277, 280–3, 285–9, 291
Marston Moor 164–5
Molesworth 296
Museum, Hendon 289, 291
Newmarket 164–5, 175, 289
Odiham 31, 41–5, 48, 61
Old Sarum 33–4, 42–3
St Athan 272–3, 277, 285, 288, 294, 296–7, 302, 306–8
Thorney Island 281
Upwood 164–6, 289
RAAF (Royal Australian Air Force) 158, 171–2, 181, 226
RCAF (Royal Canadian Air Force) 119, 172, 219
Read, F/O Henry Calvin Chambers 172
Redfern, Jim 66, 156
Rees, F/O Glyn Davidson 249
Rhine 233
Rhondda 159
Rhone Valley 248
Rhys, Richard 25, 32, 42
Rhys-Jones, Dr Ieuan 21
Rhys-Jones P/O 43
Ribbentrop 38
Rich, Flt Lt 270
Ridout, Ron 272
Riley, Edgar 'Ed' 164, 173, 182–3, 243–4, 258
River Po 251
River Rhymney 17, 35, 279

River Santerno 259, 261
River Severn 17, 27, 283, 299
Rivolto 263
RNZAF (Royal New Zealand Air Force) 181, 194–5, 204, 218, 262
Roath Coal Sidings 41
Roath Park 28
Robarts, F/O Peter 77, 79, 81, 87–8, 105, 148–9, 153
Robinson, F/O Brian Charles 262
Rolls-Royce Griffon 271
Rolls-Royce Merlin 160, 271
Romania 187–92, 194, 198–9, 201–2, 204
Rome 169
Rommel, General Erwin 92, 102, 112, 120–3, 126, 129
Ross, P/O BV 242
Royal Engineers 51, 59
Royal Navy 93, 96, 99, 113, 120, 139, 177, 282, 295, 308
Runciman, Viscount 24
Russell, Sgt 104
Russell, AS2 Sean 309
Russell, Wg Cdr Bill 175–6, 191
Russia 38, 215, 218, 223–5, 231, 233, 253, 285, 307

S

SAAF (South African Air Force) 168–9, 188–9, 204, 217, 234, 237, 265
Sadgrove, W/O 254
Sagar, Wg Cdr Phil 11, 293–5, 297, 299, 305
Salmoni, Edgar 21, 23, 28–9
Sambrook, Sgt D 196
San Benedetto 251
Sanderson, F/O Francis Andrew 204
Sandys, Duncan 286
Sardinia 143, 148–50
Sbeitla 121, 124
Scapa Flow 47
SCI 53, 73, 90
Scotland, Tom 164–6, 168, 173, 175, 177, 180–2, 236–8, 240, 242, 244–5, 247, 249, 254, 257–9, 264
Scott, Tom 280
Scott, Bill 255, 258
Scott, Flt Lt John 77, 79, 89, 91, 97–9, 100–1
Scythia 110
Sealand 28, 42
Seal, Fred 260
Self, Sgt Ivor 94, 106
Setif 111–12, 117, 153
Sfax 119, 130
Sharples, Sgt Cathy 298–9, 301
Shaw, Sgt Alec 190
Shoreham 86, 93
Sicily 126, 136, 139, 142–3, 157, 167, 171, 173–4, 289
Sidi Ahmed 133
Sidi Bou Zid 121
Sim, Alastair 62
Simpson, Flt Lt Jude 300
Simpson, F/O W 192–3, 249, 260
Sims, F/O Les 179, 200, 245–6, 252–3, 257, 264
Sims, P/O GW 113–14
Sinclair, Group Captain Laurence 104–5, 109, 112, 117, 121–4, 127, 12–32, 136
Skelton, Wg Cdr Dick 67, 71, 77–8, 82
Skinner, F/S Kenneth John 190
Sloan, P/O Dennis 97, 101, 103
Smederovo 191
Smith, Cpl Carly 306
Smith, Bill 164, 184
Smith, F/O 34, 41

Smith, F/S Donald Leonard 172
Smyth, David 77, 83–4, 93, 109, 140, 150
Smyth, Flt Lt (Llandow) 276
Snailwell 103–4
Sorbie, Sgt William Norton 113, 116–17
Souk el Khemis 115
Sousse 119, 130
South Korea 298
South Staffs 50–1, 56
South Wales Echo 22, 29, 288
Soviet Union 38, 85, 177, 245
Spain 105, 284
Sparks, F/S Clifford Harold 203
Sparrow, F/O Thomas 'Barney' 193, 202–3
Spence, Magnus 76
Spencer, Arthur 166, 176, 186, 248
Spiers, Sqn Ldr JR 282, 284
Spitfire, Supermarine 8, 18, 46, 56
RW355 270
Splott 17–18
Squire, F/S GSA 198
St Hilda's Church 276
St Marie Du Zit 133
St Mawgan 266
St Michael's and All Angels Church, Michaelston-le-Pit 291
St Trond 75
Stalag Luft III 231, 249
Stalag Luft VII 218, 226
Stalag XVIIa 219, 230–1
Stapleton, Paddy 71, 78, 89–90, 99, 105, 108, 111, 124–5, 135
Stavanger 47
Steart Flats 270
Steel, F/S EMP 198
Steua-Romana 189, 192
Stonehaven 58
Stornara 191
Strategic Defence and Security Review 293
Strathaird 107, 110
Stromboli 141
Sudden, Flt Lt Alfred 249
Sudetenland 32
Suez 306
Sugden, F/Sgt 256
Susegana 238
Sutton, Wg Cdr Harry 82, 87–8, 94, 96–7, 99, 103, 119, 132–7, 142
Sutton, Cpl Gareth 306
Swansea 273, 277, 306
Swanton Morley 82
Sylt 277
Symonds, F/O Charles 199
Szap 234
Szombathely 206, 249

T

Tactical Bomber Force 129, 132
Tafaraoui 137, 142–3
Tagliamento 238
Tangiers 108, 284
Tangmere 60, 280, 283
Taranto 172, 177, 180
Target Indicator 178–9, 182, 238, 260
Target Marking Bomb 178, 193–4, 202, 206, 254, 258, 260
Taylor, F/S GL 144–5
Taylor, F/Sgt John 113, 116–17
Tebessa 122, 125
Tebourba 134
Tedder, Air Marshal 125–6
Templer, Philip Walter 171
Terranova 149–50
Terrel, Maj Gen Fred 129
Thomas, Flt Lt Paul 308
Thomas, Lord Tonypandy George 287

Thomas-Ferrand, F/Lt Hugo 21, 25–6
Thompson, Sgt Rex 138, 140–1, 143, 145–52, 156
Thruxton 85, 87, 89–90, 92–3, 96–7, 99
Tiger Moth, de Haviland 189
Tito, Josip 243, 246
Tornado, Panavia F3 302
Tranent 63–5
Tremorfa 17, 30
Trenchard, Air Marshal Sir Hugh 24, 72
Trieste 254, 263
Tudor, Avro 275–6
Tutor, Avro 21, 25–8, 31, 36
K6103 41
Tunis 112–13, 117–19, 121, 133, 135, 142, 143
Tunisia 7, 109, 111, 120, 122, 126, 129, 134, 136, 139–40, 268, 309
Turnbull, LAC Andrew 111, 119, 127
Turner, Eric 140
Turner, F/O Thomas Joseph Rogerson 195
Twente 75, 77–8, 80–2
Twining, General Nathan 169, 248
Tyncoed Congregational Chapel 270
Tyreman, F/S Albert Edward 256

U

Udine 242–3, 256, 263
US II Corps 131–2

V

Valence 247
Vampire, De Haviland 6, 8, 274, 272–84, 286, 288–9, 291
VV196 280
VF336 279
Vechta 75, 77–8, 83
Venlo 75
Ventura 147
Verney, Flt Lt 276
Veryard, Sgt Kelly 305
Vickers 37, 117
Victoria Cross 116, 144
Vienna 218, 248
Visegrad 238
Visual Illuminator 178
Vokes, F/S Walter Richard 249

W

Wade, Flt Lt Martin (author) 295, 298
Wadi Akarit 132
Walker, Flt Lt Gavin 294
Walker, Wg Cdr Olly 11, 305–7
Wallace, Flt Lt Derek 104, 130, 140
Wallace, Lt Gordon 113–16, 120
Wallace, Westland 19
Walsh, Sgt A 112, 117
Watkins, Geoff 279
Watson, Flt Lt DQ 242
Watson LAC JAS 50, 53, 65, 72
Watson Sqn Ldr 145, 152
Wattisham 75
Weaver, Sgt 279
Weekes, Wes 'Shorty' 164, 258
Wellington, Vickers 137, 146, 155, 169, 175, 194, 199, 234
West Raynham 75–6, 78–80, 82, 274, 278
Western Mail 30, 22, 29, 33, 36
Weston-super-Mare 18
Wetzlar 218
Wheeler, Tony 168, 185, 193, 236, 249
Wheldon, F/O H 194, 206
Whitaker Road 41
Whitfield, Caroline 288

Wigston, Air Chief Marshal Sir Mike 292
Wilkins, Sgt Christopher 256
Williams, Cpl Ellie 295
Williams, P/O Gordon Melville 190
Williams, Sgt David Roland 171
Williams, Sgt Will 305
Williamson, AS1 Gemma 305
Willoughby de Broke, Lord 24
Willows, Ernest 17, 301
Willows High School 28, 301
Wills-Sandford, Sqn Ldr WR 43
Wiltshire 33, 281
Winterveldt 226–7
Withers, Sgt James Martin 172
Woodburn, LAC Leo 124
Woodward, Cpl Tom 300
Worthington, Sgt Frederick Henry 171

X
Xenia 199

Y
Yates, F/S Norman Stuart 203
Yatesbury 42
Young, F/O Thomas 112
Yugoslavia 190–1, 207–9, 211, 238, 243–4, 246, 251–2

Z
Zalaegerszeg 210

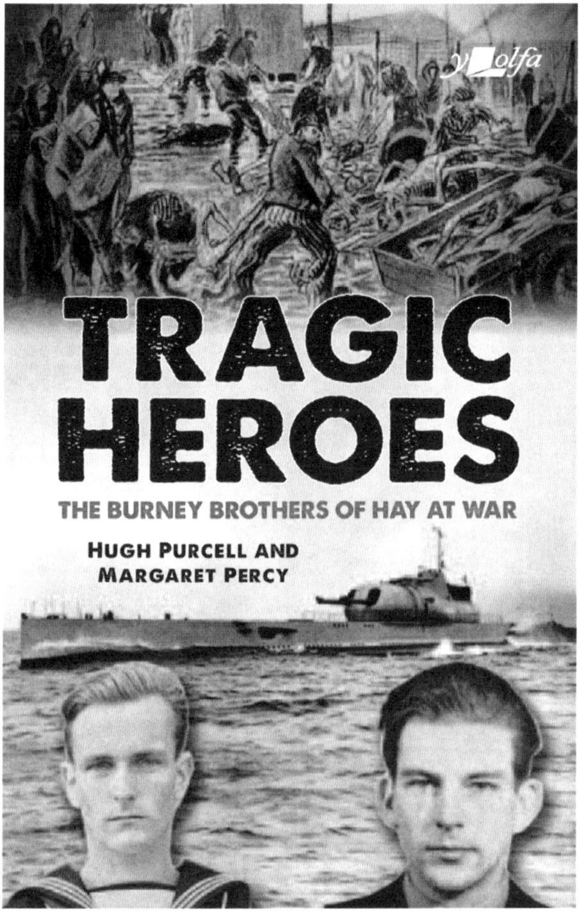
y lolfa
TRAGIC HEROES
THE BURNEY BROTHERS OF HAY AT WAR
HUGH PURCELL AND
MARGARET PERCY

£14.99

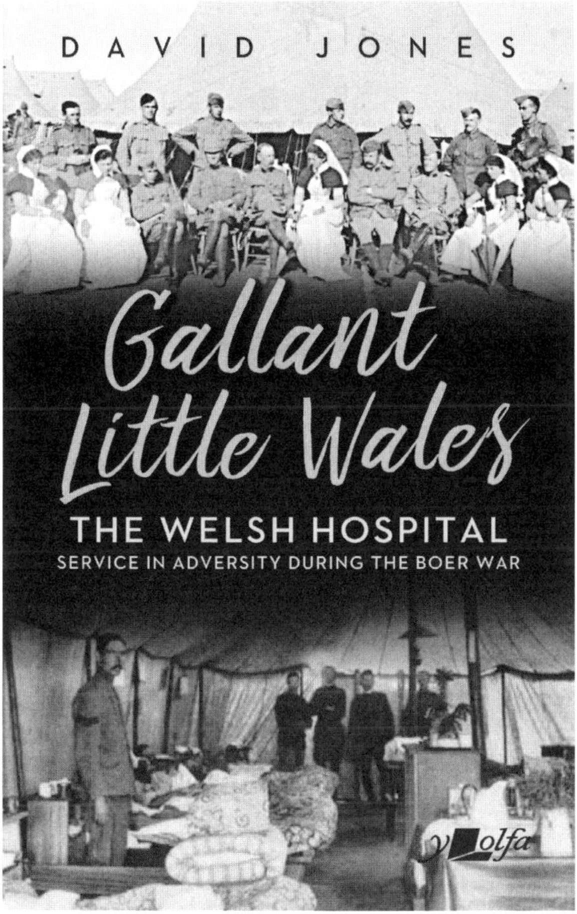
DAVID JONES
Gallant Little Wales
THE WELSH HOSPITAL
SERVICE IN ADVERSITY DURING THE BOER WAR
y lolfa

£9.99